DIGITAL INITIATION RITES

EXPERTISE

CULTURES AND
TECHNOLOGIES
OF KNOWLEDGE

EDITED BY DOMINIC BOYER

A list of titles in this series is available at cornellpress.cornell.edu

DIGITAL INITIATION RITES

Joining Anonymous in Britain

Vita Peacock

CORNELL UNIVERSITY PRESS

Ithaca and London

First published 2025 by Cornell University Press

Librarians: A CIP catalog record for this book is available from the Library of
Congress.

ISBN 9781501784446 (hardcover)
ISBN 9781501784453 (paperback)
ISBN 9781501784477 (pdf)
ISBN 9781501784460 (epub)

GPSR EU contact: Sam Thornton, Mare Nostrum Group B.V., Mauritskade 21D,
1091 GC, Amsterdam, NL, gpsr@mare-nostrum.co.uk.

To Richard, who first suggested anthropology

Contents

Diagrams

Illustrations

Tables

DIGITAL INITIATION RITES

INTRODUCTION

The attributes of liminality or of liminal *personae* ("threshold people")
are necessarily ambiguous . . . they are betwixt and between the
positions assigned and arrayed by law, custom, convention, and
ceremonial . . . Thus, liminality is frequently likened to death, to
being in the womb, to invisibility, to darkness, to bisexuality, to the
wilderness, and to an eclipse of the sun or moon.

We are presented, in such rites, with a "moment in and out of
time," and in and out of secular social structure, which reveals,
however fleetingly, some recognition . . . of a generalized social
bond that has ceased to be and has yet to be fragmented into a
multiplicity of structural ties.

Turner, *The Ritual Process* (1969, 95–96)

Some people wear this and think it's just a mask
Others wear it and know it's a symbol for the idea of change

Meme shared by YourAnonNews on 3 March 2016 (emphasis in original)

A firework whistles past my right ear. It erupts in the sky with a bang, and a passing bloom of red stars. It is one of many being set off from the stone slabs right beside me, so I move several yards away, to where the journalists and photographers have safely stationed themselves. "It's like those anarchist protests from the 70s and 80s," one of them says as we watch the combusting scene before us, "It's really got that vibe." We look toward the plinth thronged with people wearing Guy Fawkes masks, some holding signs and flags, comfortable before the cameras. They chant loudly together at intervals. Just then, another firework shoots out from the middle of the plinth with an audible *whoosh*—prompting a chorus of cheers. As I stand there, my gaze travels upward to see a purple laser dancing across the tall column, and above that, to an almost full moon glowing down on the several-thousand-strong crowd.

This is a story about how some of those attending this event—the Million Mask March—came to be here. More specifically, it is a story of a radical subjective change from being "asleep" to being "awake," and the role of digitally mediated knowledge in enabling this transition. The process takes the form of initiation. It follows ancient sequences of death and rebirth that are rendered by acting physically upon and around the human body, in order to allow for, and subsequently seal, the change that is wrought by theoretical knowledge. It is, however, a new form of initiation rite, a digital initiation rite. Unlike traditional initiation, which takes place under the auspices of ritual authorities while including recognizable rituals such as this one, a digital initiation is not a ritual as conventionally understood. A digital initiation requires rethinking how we approach the question of subjective transformation, in ways that acknowledge the profound impact of digitalization on political life.

The march begins in Trafalgar Square. It is a generous pedestrian space containing statues and fountains, and the cartographic center of London. Once known as the heart of the empire, the square was designed in the 1840s as a monument to Britain's growing imperial and industrial might. It pivots around the tall statue of naval commander Horatio Nelson, whose victory in the Napoleonic Wars cemented Britain's maritime dominance for a century.[1] From the summit of his column, Nelson surveys the wide road of Whitehall, where Britain's Prime Minister lives, and looks onward to the Houses of Parliament, the home of its political representatives. He also stands as the key coordinate in an axis designed in the early twentieth century that connects Trafalgar Square by way of a giant neoclassical arch to the house of the monarch at Buckingham Palace.[2] As the center of London, and consequently of Britain and its changing place in the world, this monument to national power has also been a place from which it has been challenged and deconstructed. Chartists. Anticolonial movements. Poll Tax riots. Political actors have gathered here to express dissent since the square's earliest existence.

But the reason people have arrived here tonight has a more elliptical genealogy, one that slides in and out of fiction. In the 1980s, a graphic novel was published called *V for Vendetta*: a thinly veiled allegory of Thatcher's Britain, written by Englishman Alan Moore.[3] In 2006 it was released as a big budget movie and told the story of an anonymous masked man likened to a modern-day Guy Fawkes who sought to overthrow the neofascist regime that had taken control of the United Kingdom. In its denouement, ordinary citizens wearing identical hats, capes, and Guy Fawkes masks assemble at the base of Nelson's Column, before making their way down Whitehall to watch the Houses of Parliament erupt in a series of explosions. On 5 November 2011, around thirty activists, who identified with the online movement Anonymous, re-enacted this scene—all wearing the same Guy Fawkes masks. Though they did not reach the top of Whitehall, and a view of their descent from the plinth was partly obscured by traffic, the short performance was recorded on film, set to rousing pop music, and uploaded onto YouTube days later.[4] "It was almost like we were creating the movie," one says.

"Somehow that just captured the imagination."[5] Within three years the march had become the annual gathering of Anonymous in Britain.[6]

This book is then simultaneously an ethnography of a new socio-technical form I consider an online cult. This term is used neither negatively nor casually. While in public discourse the term cult has taken on derogatory meanings, here I employ it as a value-neutral category, referring to a family of social phenomena that share certain characteristics.[7] Key to these are troubling experiences undergone by those who join, creating a wider sense of social ill, to which the cult offers a coherent, often totalizing solution. While outlining a transition from traditional to digital initiation, the book also signposts a transition from traditional to online cults, in which conventional aspects of this historic form must necessarily be altered. Where traditional cults are normally analog, bounded, and realized on a local scale, online cults are born-digital, open-ended, and possess the ability to reach rapidly across larger geographical scales, but potentially less ability to endure through time. Though joining cults has been similarly expressed in value-laden terms such as radicalization, programming, or brainwashing, retaining an emphasis on initiation allows not only for a more even assessment of joining Anonymous in Britain, but also more substantively, for its juxtaposition with a body of anthropological work that allows us to clarify these processes more concretely.

Alongside this anarchist edge of the march, another side is infused with jokes and play. Standing away from Nelson's Column is a young man with wavy shoulder-length hair, white cassock, and a large swinging cross who refers to himself unequivocally as Jesus. He happily poses for cameras in a perfect satire of celestial grace, and when I ask where he had traveled from that day the role is unbroken by his answer—from Jerusalem, of course—in a voice slightly muffled by his mask. Besides Jesus, a number of others have arrived in capes and robes—nods to the protagonist in *V for Vendetta*—many of which look like they have been rapidly assembled out of bedsheets. Placards are also vehicles for jokes. Many of these have been individually made with pen or paint on repurposed packaging, and alongside more earnest statements are those that teeter on a line between radical politics and humor. "ONE DAY THE POOR WILL HAVE NOTHING LEFT TO EAT BUT THE RICH," one shouts. A number of others turn on themes of love and sex.

What both sides share is a renunciation of order, what one calls "total noncompliance." It is one that extends to virtually every effort to direct the unfolding of time. Several days beforehand I accompany some of the event's key progenitors (which in this case mean the longest or most active administrators of its Facebook page) who are promoting it in more analog style by disseminating flyers and posters in central London. To one I pose the question, rather innocently, of what exactly is going to happen on the evening of November 5th. "All-out chaos," comes the pithy reply. At the time I hold open the possibility that this is a deflection. I am an ethnographer, after all, and it would be legitimate to guard social movement strategy from someone occupying a documentary role. But when the day comes it seems this was a more straightforward statement of fact. Several separate happenings have been energetically promoted on Facebook—a Mass Meditation Flashmob, a Rave Against the Police State, even a satirical dirty protest in front of the Houses of Parliament. In one way or another, all of these would yield order's dissolution through bodily methods, yet when the time arrives, even they pass into oblivion.

This is in contrast to the agents of the British state, the London Metropolitan Police, who continually seek to inscribe their own order on the march. At the beginning this is done by way of police liaison officers, affable members of the force who perambulate the crowd handing out leaflets, reminding participants of their civic obligations. Then once the event is underway, they park a small car with loudspeakers hitched to its roof straight onto the square, amplifying a sixty-second statement at five-minute intervals. In reasoned tones, and with reference to the relevant legislation, they announce a series of injunctions to all those within earshot. These are summarized again on a yellow dot matrix at the south of the square:

> Metropolitan Police Notice: Please beware of the following:
> No fires allowed or setting off of fireworks
> No amplified music no graffiti
> No climbing on buildings or monuments
> No activities that cause serious disruption to other people
> We thank you for your cooperation

In the course of the night each one of these forbidden activities comes to pass.

It is November 5th, 2014, a date known annually as "Guy Fawkes Night" or "Bonfire Night." Nominally, it is a celebration of the foiling of a Catholic plot in 1605 to assassinate the English king and destroy the Houses of Parliament, marking the day when one of the plot's leading architects, Guido or "Guy" Fawkes, was discovered beneath it with thirty-six barrels of gunpowder and a match. Yet throughout its four-century-plus history, this ritual has come to stand for contradictory things. Despite being originally instituted as a ceremony of remembrance to reinforce the Protestant hegemony, it soon took on a lively and distinct folk existence of its own, emerging as an evening for drinking, feasting, and the lighting of fires.[8] In this subaltern life, Bonfire Night evolved into an occasion for licentious behavior and score-settling, with vengeances falling along segmentary lines that often possessed an economic character. Punitive landlords had their fences removed and burned; unpopular employers had stones thrown at their homes.[9] During the particularly riotous years that followed the French Revolution, these altercations coalesced into the emergence of the "Bonfire Boys"—groups of young male artisans, laborers, and craftspeople who blackened their faces with soot, or covered their heads with cloth, and roved the streets throwing handheld fireworks and transgressing moral codes, in what was also called "Mischief Night."[10]

Early November is also the time on the British calendar most closely associated with death and transformation. The falling leaves and fading sun in this part of the Northern Hemisphere have been a point of historical continuity that lends itself to thoughts of ends and change. Hundreds of years before Guy Fawkes appeared on the stage of history, Celts had celebrated the festival of Samhain, a seasonal ritual that marked the onset of winter and considered to be a time of unusual supernatural power, when witches and fairies were abroad.[11] More recently was the medieval Christian feast of All Saints or All Souls, when prayers were said, and candles lit, for deceased friends and family members. Over and above November 5th, the vestiges of both ceremonies endure in the British ritual calendar, through the festivities of Halloween on the one hand, when people dress up in ghoulish costumes and masks, and Remembrance Sunday on November 11th on the other, when the nation commemorates its war dead. In their once religious expressions, the fires that were lit were intended to propitiate the spirits, and to protect individuals and communities from malignant forces, instigating new cycles of purification and regrowth. Mircea Eliade observes that fire rites such as these mark the end of one temporal period and the onset of another, and as such they have also been occasions for the holding of initiations around the world.[12]

"Let's march! Let's march!" An authorless chant swells from the belly of the crowd and people start taking long strides toward Whitehall. On the move, the playfulness persists. We file past a bus stop with a group of people still standing at it, and a male marcher in sportswear and trainers says to a patrician-looking older gentleman in a long coat and hat, "Looks like you'll be waiting there a while!" The latter grins at him complicitly. Within twenty minutes we are in the road in front of Parliament, in a gridlock of bodies and motionless cars, and the atmosphere becomes eerily quiet. The boisterousness that had just recently filled Trafalgar Square has somehow been muted to a hush in the presence of the building. It is almost, *almost,* as if the crowd is waiting for something, as if the explosions that conclude the respective scene in *V for Vendetta* are about to materialize. After several minutes the hush is broken by a firework going off, followed by a volley of others. The smell of smoke rises and I see a fire igniting in the distance. Then the badinage returns. The fireworks have spooked the police dogs guarding Parliament, who start barking wildly, and a policeman jokes to a small group beside me that he will throw the smallest one of them to the dogs. They laugh and smile back.

Things start to get more serious as the march moves again, this time in the direction of Buckingham Palace. Along the road that connects the two symbolically charged sites, almost every one of the crowd-controlling railings has been overturned, and people practically run toward the gates of the up-lit building. Any prospect of convergence is, however, quickly checked by a line of riot police in thick helmets, and in its wake something of a lull descends again which is quickly filled with more fireworks, and the play of lasers across the palace's façade. This time it does not last long. Again a two-word chant rises from somewhere in the crowd, "Oxford Street! Oxford Street!," and people start heading toward the border of Green Park. I bump into a collaborator helping me to document the event. "Chaos is really taking control" he says, eyes wide, before we part company once more. I move together with the marchers through the undergrowth of the park, stumbling between trees that loom up amid the darkness, and the smell and sound of fireworks going off around me. Up ahead, a man in a Guy Fawkes mask wheels a vertical black box in the shape of a coffin—the iconic face of Anonymous painted on its lid.

As the crowd pans out into the wide street of Piccadilly, the spirit of chaos becomes more materially apparent. Bins have been pulled onto their sides and sprayed with paint, and the boulevard is scattered with plastic bags and fluttering sheets of paper. By this point it is as if this human tornado dismantling the material order of the city has in its wake also dismantled the social order of difference and distance that lies between the strangers within it. Many of the drivers seated in the buses, cars, and taxis that have been frozen in the fray smile gaily at the marchers, honking their horns in percussive solidarity. The crowd's confidence surges further in this atmosphere of support. "Wake up people! No more government control!" someone cries to the bystanders. "Join us people of Britain!" yells another. As we head up the magisterial curve of Regent Street there is a sense of grandeur to the night's proceedings. "We're making history here!" a young man booms to whoever is listening, his voice resonant and strong.

When we transect Oxford Street without a second's hesitation it becomes clear that the real destination lies beyond it. The headquarters of the British Broadcasting Corporation (hereafter BBC)—Broadcasting House—glows blue up ahead, and people start running toward it. The attendant police realize this too, and they sprint to barricade the building's revolving doors. Most of the incoming crowd then swerves quickly to the right, like a river curling around a rock, and I become like flotsam flowing along with them. In the current I tumble again into someone I know, an activist in a Guy Fawkes mask who smiles silently at me, gesturing toward his smartphone which is live-streaming the scene to more than four thousand people. As we are carried along toward the back of the building, there is evidence of defacement. In *V for Vendetta*, the revolution is set in motion when people wearing Guy Fawkes masks start to secretly spray a "V" symbol across the surfaces of the public sphere. The same symbol has been sprayed in black on the glass windows of the BBC.

This encounter with Britain's broadcaster signals in many ways the climax and conclusion of the march. People continue walking for miles after that—with different swarms moving throughout the city—but all the intended destinations have now been reached. Toward eleven o'clock, my body begins to communicate its disquiet after hours of activity, and I return to Parliament Square in the tailwind of a diminutive group. Back in the ambit of Parliament the revelrous attitude returns, taking on something of the flushed and relaxed character of an after-party. A man in leggings raps on a silver snare drum, and people spontaneously start to dance. The police again attempt to control the scene by containing those present in a kettle, forming regimental lines of ones and twos.[13] Somehow, the solemnity of their efforts strikes a comic pose against the music and the reeling, and an oral rejoinder arises organically from the crowd. "Daaa daaa da daaa daa da daaa daa da daaa"—they sing the musical motif from Star Wars that accompanies the appearance of the storm troopers.

The following morning, the British press is both captivated and confounded by what has just happened. In remarkable harmony, news organizations from across the political spectrum interpret it in broad-brush terms as an "anti-capitalist" or "anti-establishment" protest, some citing Anonymous's own reasoning, namely: to oppose "mass surveillance, austerity, and infringement of human rights."[14] One of my goals, alongside other scholars of Anonymous, is to substitute this generalism that has enshrouded the phenomenon, with a detailed account of one instance of its emergence. For reasons of brevity and historical precision, I will henceforth refer to the form that Anonymous takes in Britain between 2011 and 2017 as AnonUK.[15]

The contributions of Gabriella Coleman, Jessica Beyer, Marco Deseriis, Sylvain Firer-Blaess, and Lewis Call are essential for contextualizing its emergence.[16] Each, in its own way, identifies specific ideological commitments nurtured online, that are carried through and mutated into this iteration of Anonymous. This study is the outcome of a different methodology. Following an anthropological tradition of embedding online worlds in wider sociopolitical environments, pioneered by Daniel Miller and Don Slater, and subsequently reinforced in a number of key guides, my engagement with Anonymous is led by in-person fieldwork.[17] Over the course of three years, I and other researchers spoke to more than one hundred people who associated with Anonymous, and I developed friendships and relationships with a smaller number of these. In the embedded approach, the line between "the real" and "the digital" is necessarily porous, and within and through these relationships I joined my interlocutors in online places, archiving a range of public digital materials. In AnonUK, these gathering places are overwhelmingly constituted by the hegemonic social media platforms of the time—Twitter, YouTube, and Facebook.

The embedded approach was originally developed as a critique of the cyberspace models of the 1990s that considered the Internet "a world apart."[18] The alternative it offers is to consider online presence as a banal and everyday experience, seamlessly integrated with the other domains of people's lives. For Anons— as they call themselves—this is not the case. Online experience, particularly audiovisual experience, takes on a more than normally transformative character in their accounts. The online world is a place where they encounter answers to deep questions about politics and the cosmos, that reorient their relations to them. As the epigraph reads, for those in AnonUK, the Anonymous mask is not just a mask, but a symbol for the idea of change. In the same light, the Million Mask March is not just a march, but a performance of this threshold of transformation. Many of those marching—among whom stride all the key protagonists beneath—display characteristics of what Turner called "threshold people."[19] Above we find themes of darkness and the moon, of invisibility and anonymity, death and defacement, rule-breaking, social leveling, sexuality, and play. Before historicizing this instantiation of Anonymous, and placing it in comparative context, let us consider the concept at the heart of the book.

Digital Initiation Rites

In 1909, Arnold van Gennep published a study called *Les Rites de Passage*.[20] Collecting a swathe of ethnographic material from preliterate societies and the ancient world, he discerned a remarkably consistent sequence in the rituals that changed people's relationships to the social and natural world.[21] The sequence consists of three stages. First there is a period of "separation," in which the person is symbolically sundered from their previous relationships, commonly involving some form of seclusion or "purification."[22] Then there is the "liminal" or "threshold" or period, during which the person undergoes their transition and is considered for this time a kind of nonliving person, represented as dead or prior to their own birth.[23] Finally there is a period of "incorporation," when the person returns to the community in their new guise, one frequently accompanied by commensality and celebration.[24] Van Gennep discerned this pattern across a range of rituals, including those that formalized the movements of people between different social groups, and the festivities around agricultural cycles or the movements of the sun and moon. However, the majority of the text is dedicated to so-called life-crisis rituals—ceremonies that take place around the major life changes of birth, initiation, marriage, and death—and it was these that became its main conceptual contribution. His impact on anthropology was both immediate and long-lasting, reverberating particularly through Max Gluckman and the Manchester School. Beyond anthropology, the English idiom "rite of passage" is a direct translation of Van Gennep's coinage.

As a student of the Manchester School in the 1950s, Victor Turner was vividly inspired by Van Gennep's scheme, and he focused much of his own corpus on its central "liminal" stage. With reference to his and Edith Turner's ethnographic work among the Ndembu of northwestern Zambia, Turner argued that after the rites of separation, and before those of incorporation, the ritual subject inhabited an "interstructural" stage with its own peculiar characteristics and function.[25] As its incumbents were considered nonliving persons, often symbolized as dead or in gestation, he represented it as a state of symbolic death during which they became invisible to the society in question. In this state, prevailing social and hierarchical distinctions broke down, and the relations between those undergoing transformation were often marked by camaraderie, equality, and a looseness of expression distinct from those obtained in daily life. While Van Gennep had observed patterns in the imagery and practices that had been documented during the liminal period, Turner reflected in much more depth on the creative role of this middle stage in the ritual process, which he ultimately viewed as "something positive, a generative center" during which new cultural configurations could be formed.[26]

It did not take long for Turner to see the implications of his model beyond small-scale settings. After a number of years in the United States, Turner designed the concept of the "liminoid" to refer to those social spheres in large industrial societies—notably the arts, religion, leisure, and grassroots politics—that possessed liminal characteristics, i.e., some capacity to dismantle social distinctions and generate alternatives.[27] Much like Van Gennep, Turner was instantly and lastingly influential, and the migration of his ideas stimulated countless scholars in and outside anthropology. This being said, Turner himself subsequently rowed back on some of his broader claims, expressing concern that the application of liminality to such wide-ranging phenomena "might block further analysis."[28] Although Turner's principal intellectual contribution has been to theorize those aspects of human experience beyond the constraints of structural distinction, paradoxically he always remained in a conversation about structure, to which his conceptual apparatuses were a complement, never a substitute. Liminality was a temporary state that would perennially yield to structure's reassertion. Having

RITES OF PASSAGE

RITES OF SEPARATION	THRESHOLD OR 'LIMINAL' RITES	RITES OF INCORPORATION

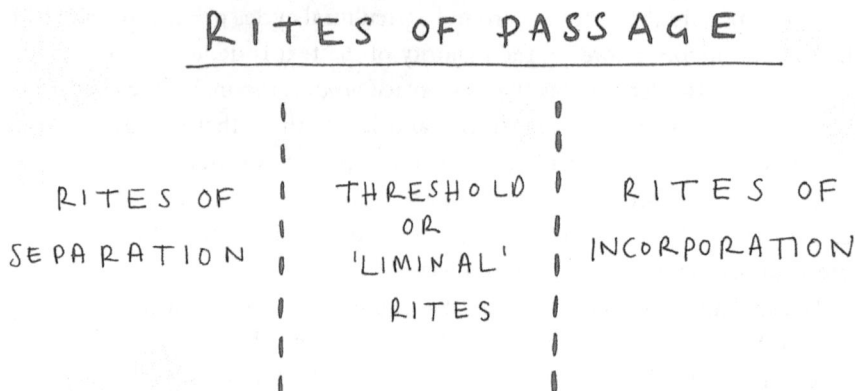

DIAGRAM 1.   Arnold van Gennep's tripartite model of rites of passage

been originally inspired by Van Gennep, Turner never departed from the latter's tripartite scheme, and consequently we can express the formulations of both thinkers with the following image (diagram 1).

Some of the uncertainties around the concept of the liminoid are resolved if we turn to the work of another Manchester School theorist—Don Handelman. Handelman expressly extends Van Gennep's and Turner's insights on liminality as an engine of transformation. Nevertheless, where the prevailing paradigm of the School was to view ritual as meaningful or functional for a surrounding social group, Handelman inverts the order of analysis. Ritual, he argues, should be examined "in its own right," not as the straightforward reflection of an outward community.[29] Handelman emphasizes the "within-ness" of ritual, its self-referentiality and self-closure, creating a spatiotemporal "pocket" into which its practitioners are more or less absorbed.[30] This inside-out perspective is critical for the argument below for two reasons. First, by exploding ritual as a cross-cultural category that must be defined by certain characteristics, Handelman greatly expands the range of phenomena that can reasonably be considered within its remit. Secondly, his approach reveals the vastly different capacities for subjective and social transformation possessed by different kinds of rituals. For Handelman, the larger the pocket, which is to say, the greater the capacity of the ritual to curve in on itself and create its own form away from its social surroundings, the greater its capacity to transform, both its subjects as well as these surroundings.

In Handelman's examples initiation rites feature frequently, and it is similarly the case that Van Gennep and Turner spend much more time examining these rites than any other life-crisis.[31] But what is initiation exactly? Even if we approach it as a phenomenon with an interior rather than exterior logic, there must still be some salients through which it can be recognized. Like other life-crisis rituals, initiation produces some change in the location of the ritual subject relative to a nexus of social and material relationships. One of the most widely documented is the puberty rite—which normally takes place during adolescence to formalize transitions from childhood to adulthood. Initiation is also key to those joining spiritual professions such as monks, shamans, mystics, or ascetics of various kinds. In addition (which is broadly the case we are looking at here) initiation ministers entry to communities across some kind of boundary, which in addition to cults include clubs, sects, secret societies, and institutions among others. There is something more specific, though, about initiation that demarcates it from other life-crisis events. While all involve acquiring experiential knowledge of the physical matters of life, initiation is the rite most closely geared toward learning things about society—particularly information that is secret and known to a minority of people. Acquiring this knowledge also arrives, to varying degrees, with a moral imperative to protect and/or reproduce what is learned therein. My operable definition is thus a sequential process of acquiring knowledge about a society that produces a moral responsibility for it.

The concept of digital initiation carries this ancient sequence of epistemic transformation into twenty-first-century forms of mediation. While acknowledging two conditions that make each of these rites irreducibly singular—the individuation of social crisis, and the personalization of online experience—they possess a broad outline of events. Digital initiands experience some form of unsettling invisibility, relative to a nexus of relationships in which they are embedded, during which they immerse themselves in a series of audiovisual experiences online through which they learn to theorize it. At an unspecified point, these digital journeys transport them to new social actors who share similar theories of the world, and the forging of these new relationships also serves to reconstitute their existing ones, making a break with the past. Beyond any social choreography, a digital initiation rite is a complex process that dismantles Van Gennep's (and Turner's) neat tripartite scheme. Nevertheless, the elements of separation, transition, and incorporation are still there, curving inward, torquing outward, in a Handelmanian dynamic that creates pockets of experience with the capacity to transform their subjects in meaningful ways. What they share most visibly with their traditional predecessors is some appearance of a rupture—what they call "waking up"—which happens within the interiority of the ritual process rather than at the start. It is this that maintains the necessity of sequence. The initiand must die to one nexus of relations before they can be reborn to another. While any attempt at formalizing human experience can, and should, be problematized, below I sketch the skeleton of these rites for heuristic purposes. As in the previous image, it is along the dotted line that relations are cut (diagram 2).

DIGITAL INITIATION RITES

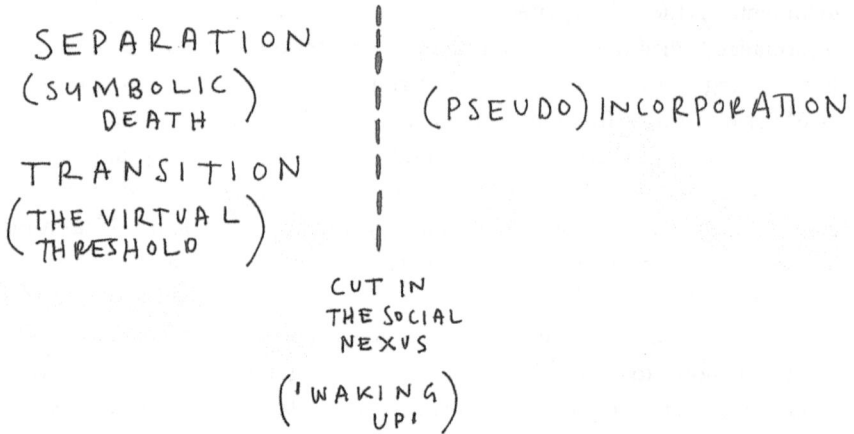

SEPARATION
(SYMBOLIC)
    DEATH

TRANSITION
(THE VIRTUAL
 THRESHOLD)

(PSEUDO) INCORPORATION

CUT IN
THE SOCIAL
NEXUS

('WAKING
   UP')

DIAGRAM 2.   Model of a digital initiation rite

A note on incorporation. In traditional initiations, the final stage of incorporation is to some degree a simplifying occasion. The ambiguities, uncertainties, forms of physical or psychological pain or discomfort that have characterized the preceding stages are concluded and resolved through food, dancing, and other forms of social union. Because of the different locations of the ritual body when initiating into an online cult, these resolutions may be less achievable in this domain. Its later stages may instead be characterized by a simulacrum of incorporation—a pseudo-incorporation—rather than the lifelong forms of social personhood that can characterize traditional rites. This is particularly the case with the more fleeting online cults. With regard to AnonUK, by 2017, and certainly by the time of writing, the principal activities of the cult have come to an end. This leaves my interlocutors in the strange position of having been digitally initiated into a cult that no longer meaningfully exists, leaving them in some ways "betwixt and between."[32] They continue to carry the knowledge acquired through their initiations—a process that is rarely, if ever, reversible—but without the collective to carry out its ambitions.

This conclusion rests a different emphasis on Turner's theory of liminality. Across his corpus, Turner consistently draws out its positive aspects, particularly in his later work which explores the creativity unleashed when structural distinctions crumble.[33] It is not that Turner does not recognize the dangers inherent in the liminal condition (which is why, he argues, these rituals are so hedged in by taboos), but he elects not to develop this underside at length.[34] The potentially destructive consequences of transitions in the subject's social life are more prominent in Van Gennep, for whom the main purpose of rites of passage is to manage and control these transitions, protecting the subject and community from harm.[35] Handelman picks up Van Gennep's point, presenting a counterpoint to Turner's optimism. "Should the liminal condition shatter its cultural buffers and instead concentrate in quotidian contexts," he says, "then its impact is searing and destructive."[36] It becomes instead a mechanism for alienation, anomie, and the dismemberment of culture itself. This critique is salutary. Handelman reminds us that as well as being creative, the experience of liminality can also be damaging—both for the subject who loses their connection to cultural life, as well as for culture which loses its connection to the internal lives of subjects.

What the practical implications are for my interlocutors I cannot with certainty say. Many go on to find different sources of meaning, although in all likelihood without the same efflorescence of communitas. What is more sure are the social implications. It could be argued that cults always exist along some spectrum of antagonism (from benign to violent) toward competing social forms. The rise of online cults, therefore, and the digital initiations into them, are logically in some degree of conflict with the abstractions that attended the rise of the nation-state—representative democracy, citizenship, and so on. Rather than being alarmed by these shifts in the social fabric, this ethnography would simply reinforce that "who cares is king." AnonUK emerged in conditions of the British state's retreat from its obligations to care existentially for its citizens, and it is not by chance that the cult sought to performatively reinject practices of care into the public sphere, as its absence became ever more pronounced. This is a good moment to turn to the prehistory of Anonymous, and the specific gestation of this offspring of it.

Who Are Anonymous?

Anonymous began life as a name taboo on the website 4chan. Anonymity by way of pseudonymity had been characteristic of online sociability since the 1990s, where people could communicate in "Internet Relay Chatrooms" using only a pseudonym or nickname.[37] When 4chan was launched in 2003—initially as a space to discuss Japanese cartoons—its administrator experimented with enforcing anonymity proper in several of the chatrooms it hosted.[38] Instead of attaching to a pseudonym, every message was attributed to the same title—"Anonymous." The experiment was not continuous and could be overruled with a technical operation called a tripcode, but before long the idea of anonymity caught on, and naming itself emerged as deviant.[39] By 2006, a group calling itself Anonymous was coordinating invasions of other virtual platforms. Beside this culture of anonymity, in which participants were compelled to subsume their own identities in a flat and homogenous collective, an intense amorality also developed, as shocking images and videos were circulated for "fun."[40] Gabriella Coleman's ethnography of these early years represents Anonymous with the image of the trickster, deities in polytheistic cultures who are known for their moral ambiguity and search for self-pleasure.[41]

It was in 2008 that this subculture on 4chan noticeably mutated and divided. Or, as Sylvain Firer-Blaess following Coleman suggests, there was a movement from "trickster" to "hero" in the characterization of Anonymous.[42] Where the former was brazenly immoral and pleasure-seeking, the latter sought moral probity and to engender social change. It is here that we find the beginnings of Anonymous as an activist force. New morally driven Anons, as they now called themselves, were especially concerned with protecting the freedom of information online. Their first success came when a wealthy American cult called the Church of Scientology sought to censure a video on YouTube, a failed effort that ultimately led to the unprecedented decision among Anons to protest the church in person outside its own centers. In February 2008 they congregated in dozens of cities across the world wearing Guy Fawkes masks—a homage to the culture of anonymity that had produced them—which became their defining emblem from that point on. A bigger climax arrived in 2010, when they undertook a large digital direct action campaign against a number of financial companies, in solidarity with the whistleblowing organization WikiLeaks.

This left-libertarian strain of Anonymous, of which AnonUK is among the last collective expressions, had its roots in the cyberutopianism of the 1990s. Social progress through technology had been part of left-wing movements throughout the twentieth century, and by its close progressive hopes began to narrow in on one technological form—the computer. The combination of expanding computational power, and the development of a way to allow computers to communicate with one another, a function then called the World Wide Web, produced a precocious interest in these devices among those with political desires. On the one hand there were computer scientists who collaborated in what became known as the "Free Culture" movement, an early counterpower to the encroaching privatization of web content, which sought to create open access alternatives.[43] On the other stood the Indigenous participants of Mexico's Zapatista rebellion in 1993, who pioneered the use of digital communications to create an international solidarity network to buttress their demand for autonomy from the Mexican state. This cyberutopianism possessed many different strands which can be gathered up into two main ideas: 1. that the free and open circulation of information online would enrich and civilize the whole of humanity; and 2. that computers enabled new forms of horizontal collaboration that would ultimately topple and replace the old hierarchical structures.[44] The former was more prevalent in technical circles, and the latter more so among other kinds of activists, but both of these ideas ultimately admixed and combusted together in the second decade of the twentieth century.

The causes of insurgency across Europe, North America, and the Arab world between 2010 and 2012 are complex and polycentric. Certain flashpoints nevertheless give the period a coherence. The first came in September 2008, when the bursting of a US mortgage bubble precipitated a collapse in the value of stock markets globally, prompting a trillion-dollar bailout of the financial sector. The second came in January 2011, when the self-immolation of a street vendor in Tunisia ignited a fire of protest across the Middle East and North Africa, against a range of authoritarian regimes. The most eventful and sustained of these was the so-called Egyptian Revolution, which centered on Cairo's Tahrir Square. Indeed, when a call was put out on an anarchist blog to "#occupywallstreet" in July 2011, it did so with the question of whether Americans were ready for their own "Tahrir moment."[45] The camp bearing this name was established in New York's Zuccotti Park in September 2011 and reflected a growing discontent across Europe and North America about the way the financial crisis was being handled by national governments. Other occupations soon cascaded in major cities across the region. In Spain, the "15-M" or "Indignados" movement—with its own sizeable occupation in Madrid—preceded Occupy Wall Street by several months.[46] Yet it was toward the latter, in naming at least, that British activists looked when founding their own camp in the churchyard of London's St. Paul's Cathedral in October 2011, with the title Occupy London.

The hopes invested in the liberatory potential of digital technologies flourished throughout this period of protest, and Anonymous was intimately involved from the start. Still reeling from the efficacy of their 2010 action, the IRC channel AnonOps, where political activities were planned, mushroomed and multiplied.[47] The projects #OpTunisia and #OpEgypt were established in January 2011 to support protesters in these states, again with the imperative of protecting the circulation of information online. In the United States, after the decision to initiate Occupy Wall Street was announced in August of that year, Anonymous released a short video on YouTube pledging to stand with the movement, and the following month published biographical and network information about a New York City police officer who had injured an Occupy activist.[48] Anonymous and Occupy, though distinct, were always closely aligned. In the UK, a small contingent emerged at the Occupy camp who identified each other by wearing Guy Fawkes masks. According to one Anon who participated, it was here that the idea to re-create the scene in *V for Vendetta* was seeded. What is certain is that this video was filmed in November 2011 while Occupy London was in situ.

One of the questions driving my fieldwork with AnonUK was as simple as it has been elusive. Who were Anonymous? Who were those behind the screens who by 2011 were beginning to assemble in this way? In-person interviews with some of the most prominent Anons of the time, in addition to discourse and image analysis, suggest the regional dominance of Britain and the United States, as well as a tilt toward youth and masculinity.[49] Beyond this, we cannot say with certainty who exactly constituted the phenomenon between 2008 and 2014. This uncertainty is due in large part to two major obstacles. The first collected around issues of security and legality. Some of the movement's activities in these years were illegal, or placed them at risk of political reprisals. All three ethnographers suggest that protecting their interlocutors was a significant motive to permit them to remain unknown.[50] The second was an important cultural coefficient. During the more exclusively online form that Anonymous took, anonymity proper was a paramount value for supporters. Deanonymizing for the purposes of ethnography would have been, in many cases, its willful transgression.

With the emergence of AnonUK, both of these dynamics shifted. Although the annual Million Mask March could include acts of law-breaking, the more quotidian activities of the cult that centered around the provision of food and clothing were often scrupulously legal. Indeed, a discourse of the law sometimes circulated in AnonUK as the civic antidote to unaccountable power. Secondly, the ethic of anonymity recognized by all analysts of Anonymous subtly mutated into its very opposite. By stepping into streets and squares as political actors, a contrary ethic of visibility emerged.[51] For the new adherents to AnonUK, declaring themselves to be Anonymous was the first step toward casting off this condition. This ideological shift finds its most vivid expression in the mask. While Anons in the pre-2011 era generally kept their faces covered, with the growth of AnonUK they became more likely to wear them on the top of their heads—what I call a thanatoid mask whose aim is to reveal, rather than conceal, the face of the wearer.[52]

These shifts allowed for an ethnographic approach that was instigated by in-person encounters. These can be broadly characterized by what Monika Büscher and John Urry call "mobile methods," in which the living with of ethnography becomes a traveling with, taking the form of "inquiries on the move."[53] In the first instance this can be read quite literally. My conversations with Anons often took place while walking or even running, while engaged in activities, or while bundled into moving cars or vans. AnonUK was a highly mobile community and I joined them in its dynamism. At the same time, mobile methods also refers to the way this approach is reconfigured through the digital in the work of Christine

Hine.[54] Traveling with, also meant gathering with, Anons online, no longer in the fringes on IRC—Facebook, YouTube, and Twitter were the main sites of interchange.[55] I also archived websites, watched Bambuser livestreams, and listened to the broadcasts of AnonUKRadio. As Hine observes, using mobile methods requires some kind of connecting thread that is able to draw together its disparate locales. At the start this thread was the iconography of Anonymous. Whatever medium this iconography took, from patches sewn onto a military jacket, to the choice of a Facebook profile picture, the appearance of the signature Guy Fawkes mask was the place where these inquiries began.

This kind of mimetic immersion can also be presented quantitatively. Fieldwork formally began in October 2013, and the first Anonymous-themed protest I attended took place in January 2014. This inaugurated an intensive first phase which lasted until the UK general election in May 2015, and a less intensive second phase which continued until the conclusion of final interviews in September 2017. Over this forty-four-month research period, I attended twenty-seven events (see **table 1**), and interviewed, with the help of collaborators, 102 people on record.[56] This does not include the larger number with whom I spoke informally. While, in the earlier stages of fieldwork, the connecting thread was the appearance of the Guy Fawkes mask, as the ethnography progressed it became the dozen actors with whom I developed deeper relationships, who themselves expressed longer commitments to Anonymous, and several of whom were central to the emergence of AnonUK as a mobilizing force. With these individuals I carried out single or multiple long-form interviews, in some cases years apart.

What follows is an anthropological rather than sociological study. Yet in view of the scarcity of demographic detail on Anonymous, it is worth offering a modest sociological portrait of these twelve people. I collectivize each category to allow for the preservation of their anonymity.

Age: Participants range between the ages of 26 and 64, with an average age of 42.

Gender: Five participants are female and seven are male.

Ethnicity and Race: Ten participants are born and raised in Britain. Of the remaining two, one is from Brazil and the other from Portugal. None of the participants would be identified as people of color, though at least one is of mixed heritage.

Religion: Two have a Catholic upbringing; two Church of England; two mixed Catholic and Church of England; one mixed Catholic-Jewish; and four are raised without religion. No information for one participant.

Labor: Seven participants are unemployed and/or receiving state benefits; one owns a small business; one works in sales; and the remaining three

are in care/service work (nursing, social care, hospitality). Only one
participant is a member of a Trade Union.

Education: Two participants possess no school qualifications; five stayed in
school until sixteen; a further five until eighteen. Of this latter group, two
completed a university degree, one as a mature student.

Politics: Parties supported by participants include the Green Party, Labour,
the Liberal Democrats, and UKIP; however seven state that they no
longer vote in parliamentary elections.

Property: Two participants own property, but only one lives in a property
they own. Eleven participants live in rented accommodations.

Some patterns can be discerned. The emphasis on youth and masculinity that
characterized its predecessor has tilted. Although not part of this research, the
increasing prevalence of women was attended in some cases by their children,
for whom its events served a pedagogical function. When considered within
the larger set of participants, a socioeconomic picture also becomes somewhat
clearer. Cited occupations include care workers, therapists, novelists, factory
workers, ex-servicemen, lorry drivers, caterers, small business owners, homeless people, university students, IT workers, and the unemployed. In aggregate it
may be deduced that these were by and large not members of a rentier class that
derives its income from ownership.

My own positionality relative to Anons was a concatenation of similarity and
difference. The latter exhibited most profoundly in our largely divergent trajectories through British institutions, particularly educational institutions. In me
this had left an affective residue of institutions as safe spaces, particularly as safe
spaces for knowledge. The experiences of Anons often veered in the opposite
direction. This introduced a productive kind of friction that was particularly
apparent online. Another aspect of digital ethnography identified by Hine is
the necessity for the ethnographer to create a presence on the online platforms
where their participants gather. Here this meant creating accounts on YouTube
and Twitter where I could follow conversations, and reciprocate in a modest way
with my own images and recordings. With regard to Facebook, I already possessed a mature profile, and this became the primary medium through which
I interacted with Anons online. Yet because of these perspectival differences,
the juxtaposition of my interlocutors with my other social and familial relations,
could create exchanges that, as John Postill puts it, were "awkward."[57] Trust met
with distrust can be like oil and water. Whatever awkwardness there may have
been, however, we shared fertile ground in common. We were concerned about
structural changes being implemented in Britain from 2010 onward, and this was
something we felt strongly enough about to take to the streets.

Deseriis argues that after 2011, new individuations of Anonymous began to rely more on local conditions and context.[58] This was the certainly the case for AnonUK. If cyberutopianism and the capacity to wake people up online was the carrot that inspired people to join Anonymous, then the UK government policy called "austerity" was the stick. The short history of austerity begins with the financial crisis of 2008. Asset prices collapsed as well as tax receipts from the financial sector, which constituted some 25% of all fiscal revenues. In response Prime Minister Gordon Brown guaranteed 40% of Britain's GDP to rescue the banking system, substantially increasing the national debt.[59] However in 2010, when a Conservative-Liberal Democrat coalition came to power ending thirteen years of Labour rule, it did so with a fully formed narrative of how this debt was the consequence, not of financial volatility (and malpractice), but of excessive government spending on the domestic organs of the British state.[60] In contrast to countries such as Greece where austerity was imposed from without, in Britain the decision to cut public expenditure was imposed from within, which meant that by 2015, an estimated £64 billion was wiped off the revenues of central and local government.[61]

The effects of this policy on Britain were immense. Among others, funding was significantly reduced for the institutions of care (social care, the benefits system, the National Health Service); for institutions providing public access to knowledge (universities, schools, libraries); while some mechanisms of social opportunity were abolished entirely (such as the Educational Maintenance Allowance, Legal Aid, and the Future Jobs Fund).[62] It is important to stress that despite the political rhetoric of shared sacrifice, austerity was never evenly imposed. Around 50% of all reductions fell on benefits and local government, despite these constituting just 27% of all expenditure.[63] It was thus disproportionately weighted toward those who relied on the state for income, housing, and social care.[64] At the same time, other government departments were not only exempt from austerity, but in fact saw their budgets substantially increased. Foreign Aid rose by 21% and the Treasury, the Cabinet, and Quangos saw a 242% rise in their own revenue streams.[65] The targeting of austerity had predictably targeted effects. The use of Food Banks (centers for the free distribution of food and necessities to those on low incomes) grew exponentially between 2009 and 2015, while mortality rates also rose significantly, most of which was attributed to a shortage of nurses.[66] In 2018 the UK was investigated by the UN Human Rights Council, which found that despite being the world's fifth largest economy, some fourteen million Britons were now living in poverty. As of April 2025, this figure is roughly the same.

Austerity was presented as a straightforward matter of economic expediency. Numerous scholars have, however, observed the continuity of austerity with the commitments of previous Conservative governments. Margaret Thatcher's policy turn in the 1980s also combined benefits suppression with regressive taxation and the nurturing of financial services.[67] Even further back, Stanley Baldwin's government of the 1920s likewise rejected investment in public works programs as a solution to the postwar slump, famously maintaining just before the Wall Street Crash of 1929 that the state was "crowding out" the private sector.[68] In 2013 Prime Minister David Cameron rehearsed these long-held Conservative views about the state, when he acknowledged that what his government sought was to make the latter "leaner, not just now, but permanently."[69] With the view of hindsight, the class politics that was always at the heart of the austerity agenda becomes more obvious. In an anthropological volume on the global life of austerity, Theodoros Rakopoulos and his coauthors argue that wherever it was imposed, austerity was a project of social change designed to protect the hegemony of creditors (people and organizations who own assets and investments), a set of actors with whom the Conservative Party has held a historic alliance.[70] As poverty rose the value of assets rose too, their owners directly benefiting from the deflationary economy produced by Britain's slowest recovery on record.[71] Don Kalb goes one further to remark that of all the austerity policies pursued globally, those in Britain were "the least embarrassed," combining them with tax cuts for the highest earners.[72] In the 2010s, Britain was a place where the cultural hegemony of creditors was so pronounced, they were able to dominate the consensus on the ethics of public financing with astonishing success.

In these local conditions, AnonUK emerges as part of a wider left-wing mobilization against this consensus. Anons enthusiastically join protests organized by anti-austerity groups—not only Occupy but others such as UK Uncut, the People's Assembly, and the Trade Union Congress. In practical terms the British state is their primary focus, and virtually every event chronicled below involves some critical encounter with it. Nevertheless, one of the things that constitutes AnonUK as a cult rather than a political movement per se is that despite the class politics of austerity and its peculiar national character, ideologically speaking, Anons did not see themselves in the terms of class or nation. Instead, waking up furnished them with a triangular view of human society, one that I formalize with the pronouns that most commonly attend each segment. While such distinctions are often implicit in cult studies, I visualize them here to streamline the argument later on (diagram 3).

Waking up initiates them into the loose community of the "We," commensurate with Anonymous. Anonymous had introduced itself with repeated use of this pronoun several years earlier, when it announced itself on the global stage with a promise of revenge. With the maturation of AnonUK, the two other pronouns appear. "They" are a small but powerful group of global actors that includes those who hold institutional and economic power in Britain, while "You" are other ordinary members of the (implicitly British) public, who may be strangers,

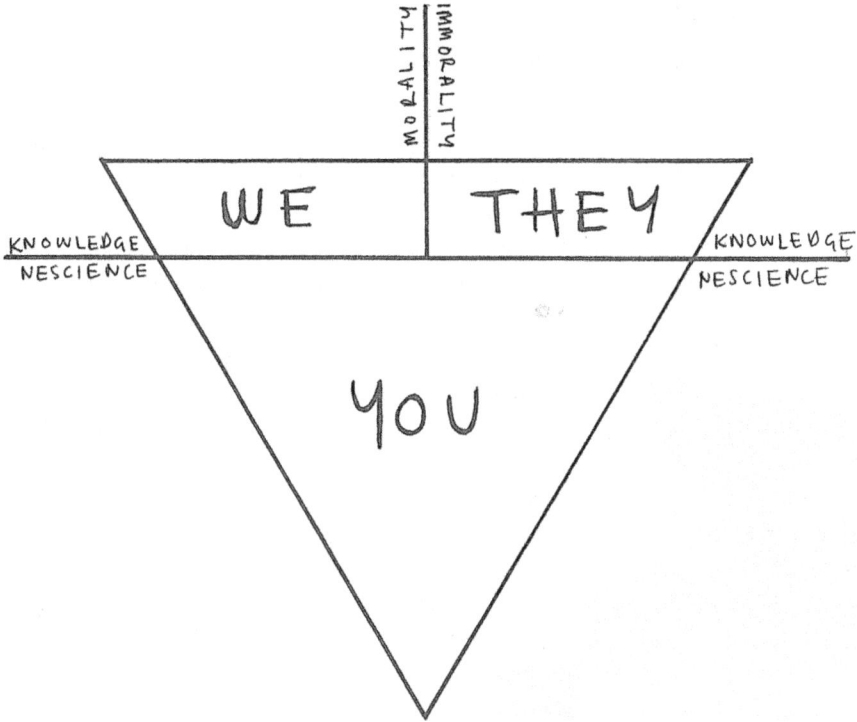

DIAGRAM 3.   Model of social segmentation rendered by a digital initiation rite

friends, or family members. The first axis that segments the system is knowledge. Here this means knowledge of a specific doctrinal kind, namely knowledge of a deadly war (sometimes called the Depopulation Agenda) that They are waging against We and You through a variety of surreptitious methods. Following a long discursive tradition, the possession of knowledge or nescience is represented through the idiom of consciousness. The second axis is morality. While We are commensurate with the good and They with evil, the You occupy an ambiguous position and Anons' feelings toward them can also be mixed. Despite this, much of their political energy is dedicated to awakening the You who constitute their primary audience, and rehearses their own travel across these two categories. This image is deliberately asymmetrical. The early ideology of Anonymous was strictly equalitarian, and though this persists into AnonUK, there is a tacit hierarchy of We and They over You which turns on the presence or absence of knowledge. Having introduced this community and my ethnographic engagements with them, and traversed the main theoretical landscape of the text, I turn to the comparative context within which it speaks.

Online Cults

What makes something a cult? The genealogy of the answer begins with Max Weber's distinction, advanced by his student Ernst Troeltsch, between a church and a sect.[73] A church is a form of institutionalized religion that people are born into, while a sect is a breakaway group with voluntary membership, that asserts itself as an act of protest against this and/or the wider society in which it is located. Central to this schism is the concept of salvation. Members of sects often join in the wake of highly charged emotional experiences, sharing a conviction that there is an evil in the world to which the sect offers the solution. The concept of a cult has developed from this concept of a sect, and the consideration of varieties of religious forms that tend to share a Christian millenarian bent. In anthropology, interest in cults began to grow from the 1950s onward, particularly through the cargo cults of Melanesia and the nativistic cults of the Americas, viewed as a cosmological response to abject political conditions.[74] Sociologists, meanwhile, were concerned with the growth of cults in Europe and North America, in the wake of the 1960s counterculture and the demise of institutionalized religion.[75] In sociology, cults are often thought of as the persistent recurrence of a religious impulse in the context of mass secularization.

Rather than tying AnonUK to one of the existing typologies, I offer an inexhaustive set of salients to suggest it should be considered within the family of phenomena called cults. The first is a shared positionality among participants that sociologists have called "relative-deprivation."[76] This is the notion that members are not necessarily socioeconomically deprived (though they may be), but are deprived relative to historical expectations. The politics of austerity in Britain produced a gap in what the British welfare state was becoming vis-à-vis what it was historically considered to be, and AnonUK was one of the entities that occupied this space. More importantly, theirs was not an abstract critique but something many of them experienced directly, as detailed in chapter 1. The second is a shared and to some extent totalizing vision of the ills of the world and the path to its salvation, addressed in chapters 2 and 3. Then there are the powerful affective bonds between members—which should be apparent throughout, and to which I as ethnographer was far from immune—as well as a proselytizing relation between members and nonmembers, described in chapter 7. Lastly cults are often bound by some form of esoteric knowledge, a theme I return to continuously, but with particular emphasis in chapter 5.

AnonUK also differs in key ways that may or may not be representative of how the cult form is changing under the impact of digitalization. The most stark is its horizontal or "leaderless" form. Unlike many twentieth-century cults that were hierarchical in orientation and hinged on the authority of a prophet or leader,

AnonUK had a decentralized structure with no explicit acceptance of any coherent source of authority.[77] As is often the case however, one of its consequences was to create avenues for tacit or opaque forms of authority to extend themselves—as documented in the discussion of digital witnessing in chapter 2.[78] Connected to its decentrality is an emphasis on self-organization rather than bureaucratization. This leaned heavily on social media as a means to share information and raise resources which, while having a number of episodic successes—chronicled in chapters 4, 6, and 8—ultimately worked against the cult's ability to endure through time. AnonUK remained, and effectively dissolved, in what Francesco Alberoni named the "nascent state."[79] While these founding stages of political or religious societies release an efflorescence of communitas, they must give way to more robust internal structures or be engulfed by those beyond them.[80]

The sociological interest in cults expanded in the context of the cult scare of the 1970s, and the rise of so-called "destructive cults."[81] Since then the term has been an emotive one that can produce an alarmed response. In the twenty-first century, much of the scholarly and popular interest in cults has dissipated, and the aura of alarm arguably transferred to a new object—conspiracy theorists.[82] Although Anons share theories of the world that scholars of conspiracy theory would recognize as falling within its remit, I avoid the term as one which is heavily value-laden and of uncertain utility in this context.[83] The experiences that led people to join AnonUK can be understood as a diagnostic of power, and the theories that follow an attempt to diagnose power, albeit through methods most scholars would not accept.[84] Approaching the phenomenon anthropologically I employ the term "vernacular theory." This serves firstly to excise it from the social panic in which the phenomenon is increasingly enveloped, as well as inviting greater attention to the regional conditions that give rise to their own theorizations and effects. I should perhaps add that it is not that the rise of vernacular theorizing with and through online cults is not concerning for democratic societies—particularly in their explicit link to stochastic violence and failed insurrections in Germany and the United States—but that alarm is an affect that can readily lend itself to reactionary outcomes. For this reason it is one that, here as elsewhere, I endeavor to avoid. This is undoubtedly easier because AnonUK was a nonviolent movement that fell down on the left rather than the right side of the political spectrum.

Just weeks after my fieldwork ended, however, another online community of "Anons" was birthed on 4chan.[85] On 29 October 2017 an anonymous post appeared on the /pol/ message board ascribed to someone calling themselves "Q." With a tone of authority and written in a style suggestive of a person or group familiar with Chan culture, thus began a series of prophetic statements about unfolding political events that soon developed an avid following. In contrast to the leftist sympathies of AnonUK, this new collective known as QAnon became a focal point for the espousal of authoritarian and neofascist ideologies, and the legitimation of violence. Scholars and commentators generally concur that this ideological schism on the platform emerged around 2015–16, as 4chan became subject to "polarization effects," when the absence of those holding moderate views reinforces extreme positions.[86] By this point left-leaning Anons had left the message board for other venues, and far right culture began to grow in confidence, particularly on the /pol/ board. Although they were on different sides of the political spectrum, it is important to recognize that earlier forms of Anonymous share with QAnon a transgressive antimoral style that 4chan nurtures, long known to hold the potential for dangerous forms of politics.[87] QAnon could be seen as this potential brought to fruition. Yet it is only one of other global examples of how far right-wing culture online is now characterized by what Sahana Udupa calls "fun," just as Anonymous was first described.[88]

QAnon is often understood to have the qualities of a cult and has further stimulated an active debate around radicalization online. This literature excels in its examination of the linguistic, visual, and audiovisual means through which people arrive at radically different political convictions through new media, particularly the role of simple and compelling narratives.[89] For methodological reasons, there is less empirical focus on the actors themselves. Indeed, some scholars of online extremism have noted the dearth of ethnographic investigation in this field partly because, as with waking up, the phenomenon is only known to scholars largely after it has already occurred.[90] As a consequence, explanations for vulnerability to these processes have a strong cognitive orientation, with an emphasis on the language of feeling—in which this feeling is understood as one of alienation or anomie.[91] While concurring with the importance of feeling, here I offer a response that places it fully in the context of the embodied subject, not just cognitive processes. The accounts of those inside AnonUK, and my engagements with them, suggest the central role of embodied experience before, during, and after waking up, in providing the possibility for life-altering epistemic change. Lastly then, I return to anthropology.

Anthropological Comparison

This book is an exercise in, and outcome of, anthropological comparison.[92] I situate the experiences of Anons in direct conversation with some of the most enduring anthropologies of initiation. This comparison is not one of simile. Becoming a member of this online cult was not *like* initiation, I submit, it *was* initiation, albeit not in a form that anthropologists would immediately recognize as being.[93] Underpinning this is a more widespread acceptance among anthropologists that the digital transformation has so fundamentally altered the landscape of human relation, that social phenomena which preexisted it must be reconsidered and reconfigured in its wake.[94] I did not set out to study digital initiation into an online cult, but some of the political movements that were arising in Britain in response to austerity, and the role of digital media in constituting them. Instead this subject continually imposed itself, as I traveled into the anthropology of initiation over several years. What follows is consequently not intended to be a definitive statement on the subject, but the presentation of a perspective hitherto unexplored, on the dynamics of radical political change online.

In the early studies, initiation was viewed in explicitly evolutionary terms. While James Frazer saw initiation as "the central mystery of primitive society," he also believed that it would ultimately be superseded by modern science and disappear altogether.[95] Even in the 1950s, this prediction persists into the work of Mircea Eliade, who stated that in places such as Britain the institution was "practically nonexistent."[96] Behind these predictions lay not only the brutal cultural hierarchy of evolutionism, rendered ideologically self-evident by the technological dominance of the northern empires, but also the notion that initiation was an exclusively religious phenomenon that would have no place in societies governed by scientific rationalism. The case of AnonUK profoundly contradicts the future that Frazer and Eliade envisaged. While the truth claims of scientific rationalism are more powerful than ever, here they are unmoored from its forms of social organization and its methods, and overlay a phenomenon that resembles a new kind of secularized religion. Moreover, initiatory processes have not disappeared with the advent of modern technologies, but in fact have changed and even possibly expanded with and through them.

Very different overarching assumptions are not the only risks of bringing this older literature to bear on a relatively nascent phenomenon. Colonial forms of representation frequently flatten ethnographic subjects into homogenous collectives, with little or no characterization nor direct quotation. There are some who would go as far as arguing that this literature is so corrupted by the colonial conditions of its production, that it should be disregarded in its entirety. I sympathize with this position. Whenever I came across initiation masks in the ethnological museums of Europe, my affective response was perpetually one of deep ambivalence. While I was fascinated and stimulated by these objects, and what they communicated about the common history of the human, another part of me was nauseated by their inertia and exposure behind the vitrines. This was especially so, given how sacred and/or secret many of these masks were originally intended to be.

This being said, I offer that the ethical and intellectual gains advanced by these texts outweigh the risks for the following reasons. Many of the initiation rites discussed below no longer exist, or exist in a much altered form. Like other rituals and practices that preceded the break of colonial modernity, traditional initiation was a major casualty of its arrival. Through the advance of Christianity in some directions, Islam in others, the spread of European-style education—as well as a deliberate and sustained suppression of these rites—many faded or disappeared in the decades that followed.[97] Given this history of actual and symbolic violence, silencing these older ethnographies constitutes a double erasure, as these documentary residues are likewise lost to posterity. An ethnography is not only the product of its author, but also the product of all those dozens of often nameless interlocutors who give their time and energy, to help the ethnographer understand activities that matter to them.[98] By animating this comparison I seek to reconstitute these ethnographic subjects as historical actors, capable of producing anthropological theory in the twenty-first century. Fred Myers asks, in a discussion around this subject, "Are there forms of sociality evidenced in the earlier works that might illuminate current situations?"[99] My answer to this question is yes.[100]

Employing historical studies of initiation is not the only engagement with the past. In order to vernacularize AnonUK, and assess some of the significant changes in the transition from traditional to digital initiation, I make substantial use of history. This begins with the oral histories which relay biographical sequences of waking up. Later there is more emphasis on the history of place. I explore some of the urban landscapes in which Anons congregate, in shallow and deep historical ways, to trace some of the political, economic, and religious trajectories that have rendered this cultural formation. Indeed as one of the few places in Europe where feudal relations were never truly interrupted, it could be argued that deeper forms of history have a particular analytical salience in Britain.[101] I also draw at certain points on the history of ideas, political and theological, that have informed some of Anons' primary assumptions. Lastly I integrate cultural history, particularly two films made by the Wachowski sisters—*V for Vendetta* and *The Matrix*—which in their impact demonstrate the importance of audiovisual artifacts accessed across global media flows. Given that AnonUK no longer meaningfully exists, this ethnography has also passed into history. I hope it is one that may illuminate the futures into which we walk.

Part I
DEATH

BODY

To die is to be initiated.

Plato (cited in Eliade 1995, 111)

Among the Orokaiva of Papua New Guinea, the puberty ritual for girls and boys begins with high drama. Without warning, elders wearing elaborate masks made from bird's feathers and pigs' tusks pour into the village from the surrounding bush, making loud sounds like bird calls and shouting that they are the spirits of the ancestors. They chase the children who are going to be initiated around the village, while the former run terrified and their parents plead with the intruders not to "kill" their offspring.[1] Finally the children are herded together onto a central platform, similar to those on which pigs are slaughtered during other village ceremonies. Their heads are covered with capes and thus blindfolded, the initiands are removed to an isolated hut in the bush away from their relatives. Once inside the hut, they are forbidden from eating normal foods, nor permitted to wash or speak. They are told they have become the spirits of the dead.

Anons tell stories of separation. Whether slow-burning or eventful, whether in the recent or the distant past, they recall experiences in which their position in a nexus of relationships broke down, during which they felt that they no longer existed relative to these relationships. In these accounts, this break is foundational to becoming Anon. Because all these experiences take place in the course of daily life, that is, without being overseen by any singular or coherent ritual authority, each of these stories possesses its own idiosyncratic character and emphasis. Nevertheless, all contain notes of what Victor Turner called symbolic death—often drawing explicitly on the imagery of death itself.[2]

The stories rest on a spectrum of intensity. Some only describe an episode of severe discomfort or societal critique. Those that are relayed the most intricately and most passionately, however, involve some profound experience of the body: whether of an involuntary seclusion, internal dysfunction, or where its boundaries have been involuntarily breached. This chapter relays three of these stories at length, which are also indexical of others untold, while shorter stories arrive in subsequent chapters. When considered together, it is bodily experiences that appear to be the most transformative. I am though not proposing a kind of biological determinism. Simply acting upon or around a human body does not necessarily sever it from its position within a nexus of existing social ties. Instead, it is when these embodied experiences are embedded within some broader social failure of care that a powerful pathway into AnonUK is opened up. In fact in their narration, it is not the (sometimes dramatic) events around the body that carry the most emotional weight, but the absence of physical or spiritual care at the very moment when care was needed.

The premise of all rites of passage is that material acts of separation are required to instigate the rite, temporarily removing the ritual subject from personhood. The history of initiation offers countless examples of this premise, ranging from mild to highly extreme. There are the painless practices of hair cutting and shaving, the painting of the skin, or the covering of the body or the eyes.[3] As with the Orokaiva puberty ritual above, this is frequently combined with some deliberate physical removal—when bodies are rendered literally invisible to become symbolically so. This could be some designated or specially constructed space such as an initiation hut, monastery, or lodge, or achieved, particularly for those entering spiritual professions, through a long period of roaming far beyond the initiating community.[4] Rites of separation have also involved presumably excruciating mutilations—piercing, tooth extraction, circumcision, the cutting off of the earlobe or the tip of a little finger.[5] Where initiation involves mutilation, the healed scar often becomes a badge of belonging, signaling the permanence of the change made. In Germany's nineteenth-century fencing fraternities, successful initiation was marked by a scar on the face, often to the left, visible to strangers and intimates alike in perpetuity.[6]

Explanations for how people arrive at extreme positions or theories online generally stress the role of cognitive experience. In studies of right-wing radicalization, scholars and commentators cite a loss of control and status anxiety among the (often) young men subscribing to these positions.[7] Alice Marwick and Rebecca Lewis draw on Durkheim's concept of anomie to describe the individual psychology of a failure to achieve what has been historically anticipated, and a resulting fragmentation of identity and loss of purpose.[8] The implication is that these feelings are sufficient to produce what Anna Kruglova calls a "cognitive opening," through which extreme interpretations for this angst are able to find a foothold.[9] Journalist James Ball goes the furthest into this idea of cognitive susceptibility, representing QAnon extensively as a virus that anyone online could theoretically catch.[10] This being said, several of these authors, who work with discursive methodologies, recognize the need for complementary forms of research that are able to describe the life situations of theorists more fully.[11]

Below I summarize the events that led three men—Patrick, Saul, and Gregor—to become regular participants in AnonUK. While the narratives of Patrick and Gregor both involve embodied experiences, the first of being physically assaulted, and the second of a prolonged banishment from the world of work, Saul's story is of a deep critique that comes about through seeing the effects of austerity policy on someone else. Significantly, while Patrick and Gregor become highly committed supporters of AnonUK, Saul rapidly becomes disenchanted. Seeing these in juxtaposition suggests that a feeling of alienation or anomie is not sufficient to bind people at length to the projects of AnonUK. As with initiatory passage, it is social practices around bodies that provides the more efficacious precursor for membership. Here I relay these events in the manner in which they were relayed to me, which is to say dramatically, and in this way expressive of how AnonUK grew out of a story, that was based on a story, that was based on four centuries of vernacular stories.

In *V for Vendetta*, scarring, both physical and psychological, is an important part of the plot. In the graphic novel, the back story of V begins with his imprisonment in a scientific facility where he is subjected to involuntary biological testing, the name "V" coming from the Roman numeral for the number five, which is his bureaucratic identifier. V's escape comes when he sets the facility on fire, grossly disfiguring his own face with burns in the process. One of the reasons that V wears a mask in the first place, in both the novel and the film, is to cover these unsightly scars, which also provide a physical memory of his subjection. But in the process of covering them, he assumes a new persona of resistance. This dynamic, from trauma to resistance, appears in some form in every pathway to AnonUK, and in more than one are the scars physically visible.

Attack

Patrick leans his back against a tombstone and starts talking. He was born on New Year's Eve in the 1970s in eastern England. His father was an ex-football player and member of one of London's biggest gangs, his mother an Irish Catholic with Romany roots. She had been on antidepressants and in and out of mental health institutions while he was growing up, so as a child he always had a lot of contact with agents of the state—social services, police, doctors, caseworkers—which made him, he says, "very anti-authority."[12] At sixteen he left home and school and joined the army. He was enriched by the experience, he says, recalling it at length, but because of his already developed anti-authoritarian instincts he was discharged after a year, and from there his life took a different path. He started going to illegal raves in warehouses and fields, and he was soon organizing some of his own in an old nuclear base. "I was a raver," he recalls, "always going on about the next event." The 1990s British free party scene was part of a wider subculture of civil disobedience and antifascism that Patrick similarly blended in with. He would bring along his own sound system to Reclaim the Streets, spontaneous occupations of roadways that combined environmental politics with rave culture. By the early 2000s, Patrick had formalized his love of nightlife into owning a bar, and it was just after dropping in there for a midnight drink that everything changed.

Time was sliced for Pat in 2008. The same year that supporters of Anonymous first appeared on the streets of Britain, and stock markets tumbled across the world, instigating a period of global unrest, Patrick experienced a decisive turn in his own life-course. It is one of those moments where biography and world history fuse in a flash that combines them so tightly that afterward it is impossible to pull them apart. He had had fun that night in his bar. He romanced one of his cousin's friends and had just put her safely into a taxi "doing the nice thing and all that." Shortly after the car door slammed, he saw a male security guard nearby behaving aggressively toward another woman, and in the chivalrous mood he was in he intervened. "Come on mate," he said to him, "Don't hit a woman. If you're going to fight someone, fight a man." "What, like you?" the security guard replied, before quickly disappearing.

Ten minutes later the man returned with a group and time slowed to a still. Having been born on a night of revelry, something told Patrick he would die on one too—a thick thread of determinism running through his long tale. "He wants to have a word with me, so I go over, but then click, something ain't right . . . I knew I was going to be dead that night as soon as he grabbed hold of me." Patrick sustained injuries from knives, bottles, and boots in the near fatal attack, but in his recollection of the event, it is the social abandonment that stings more keenly than the memory of his injuries. The first came when his friends ("the so-called hard men of this city") jumped in taxis and left. The second came when the police arrived. Patrick believes that because one of his friends was known for criminal behavior, the police intentionally delayed the arrival of the ambulance. "So I was left there by the police to die on the pavement . . . They really wanted me dead."

But the ambulance did arrive, and Patrick was treated in the Accident and Emergency ward of his local NHS hospital. Straightaway the doctors injected him with something for the pain as they began surgery. He remembers them saying, "How the hell are you alive? This is a miracle" before he stepped into a shaft of light. Patrick entered a long tunnel with a woman standing at its mouth. She was tall and pale and had white blonde hair and exuded a powerful presence over him. "Some would say Jesus spoke to them or an angel spoke to them or God came down and spoke to them. But it was a woman and she was wearing an Anonymous mask." She had a message for him. "To become Anonymous . . . To fight a war . . . To lead a revolution," not as a leader per se, but "keep on doing what you're doing, start being forthright." He describes this without any sense of contradiction as the night he "died."

Prophetic visions—sometimes accompanying physical traumas—are among the normal inaugurators of new cult forms. In Eileen Barker's influential study of the Sun Myung Moon Unification Church in the 1970s and 80s, she describes the physical tortures and ecstatic revelations of its founder, Moon.[13] In Moon's first vision, on Easter Day 1936, Jesus appeared to him and revealed that he had been chosen to establish a new kingdom of heaven on earth. It is clear that both Moon's and Patrick's accounts of visionary experience bear a not inconsequential resemblance to the conversion of Paul to Christianity on the road to Damascus, where he was blinded by a brilliant flash of light before being instructed by Jesus. In the case of Moon (just as with Paul), these and later visions were collected into a written text which became a central source for future adherents.[14] But the social afterlife of visions does not need to be logocentric. In the cargo cults of Melanesia and Polynesia, visions relayed by a variety of prophets yielded collective rites that, followers were told, would bring European goods on shore.[15] Some peculiar aspects of Patrick's tale can be brought out. The first is its differing relationship to authority. Embodying this, the apparition is not male (who might then be interpreted as V), but female, evoking for Patrick a kind of decentralized blessing, an instruction to become a leader who is not a leader.[16] Entwined with this is the absence of prescription. She relays a message without a message, encouraging the doing of good works, but in which the nature of the good remains undefined. The only certainty is that he must become Anonymous.

It is Patrick who selects the cemetery as the site for our discussion. He is unfazed, perhaps even comforted, by the presence of human remains beneath us, describing it to the warden who moves us on as "somewhere nice and tranquil." As he continues his story in the dappled shade of a nearby park, the conversation becomes more mystical, and it begins to have a tangible effect on him. He describes his capacity to travel into other people's dreams, and how this allows him to predict the future. Just as the woman wearing the Anonymous mask in his vision passed on a message, he also receives other kinds of messages. In fact "it is happening right now," he says, "look, can you see them?" He points to the blonde hairs on his right arm, all of which are now standing up. "That's how I get my messages." What is the message, I ask, curious. "I don't know," he replies, "but I will follow it up. It means something about today." Another consistent aspect of visionary experience is the desire to communicate it. It is not confirmed whether or not Patrick has told anyone else about this particular experience in the hospital (as Barker reports of those who become "Moonies," people in secular Britain tend to keep their visions to themselves) but the appearance of goosebumps suggests the stimulation its articulation is conferring.[17]

At a food distribution event in his home town, Patrick is interviewed by a local radio station about how he got into activism. In a far shorter and more sanitized version of the story he tells me, the visionary and mystical aspects fall away, as do the gruesome details of the attack. Instead, what remains is entirely social, as he conveys the sensation of abandonment.

> It's something you find yourself falling into, by the way you are treated by society, or treated by any support group who are meant to give you support, but they're not actually there to give you support, so at the end of the day you find out if you want help, you have to start with yourself, and try to help others, and it seems to spread from there.[18]

The salvation narratives that Anonymous offers allow him to turn this sensation inside out.

Hunger

Saul is seated on the slabs of Trafalgar Square. He faces the onlookers and tourists wearing a sweatshirt with an anticapitalist slogan on it, his back turned to Downing Street and the Houses of Parliament. Gradually, men and women sit down in the square on either side of him, forming a line of a dozen or so which curves to a J at one end. Saul is leading an open public meditation, and he begins the ritual with a short speech about how meditation offers its practitioners the capacity for "awareness." In a deliberative gesture, he then lowers the Guy Fawkes mask which had been resting on the top of his head, to a new position over his face, prompting several others around him with masks of their own to do likewise. With one hand he cups a small metal singing bowl—a common Buddhist meditation aid—while the other slowly circles its rim with a stub of cylindrical wood. Almost imperceptibly, the bowl begins to emit an audible thrumming sound, broadcasting the beginning of twenty minutes of public quiet.

Three months later, Saul walks the coast of the English seaside town where he lives and works. As the North Sea pours over the flat rocks, and the wind thrashes furiously around him, he recalls the sequence of events that brought him to the square that day. He first came across Anonymous online. Initially, he confesses, he found the mask rather "menacing." A friend of his mentioned it at his local meditation group and his immediate reaction was, "It's not my bag." But he continued to look into it. In spite of what he saw as a slightly feigned theatrics—particularly the videos of a masked figure making announcements in a computerized voice—he felt that there was a genuine sense of empowerment underlying it. Increasingly, he started to view Anonymous as highly aligned with his own values, particularly the spiritual principles he had nurtured through several years of meditation.

> The whole ethos of Anonymous is really, it's not about me, it's not about trying to aggrandize oneself. The mask is great, because actually it says, "We're not here to be rewarded or to raise our profiles." It's a sense of, "We're here for the principle," as opposed to a furthering of my own ego.

Then he came across a music video put together by Eminem and several other American rappers which had the aim of promoting a crowdsourced wave of political action led by supporters of Anonymous.[19] Saul felt something inside him shift. Doing activism online is "all well and good," he says, "but the whole point is you need to materialize it in some way." It was this video in particular that prompted him to make contact with other Anons and contribute to this idea for mass mobilization with the mediation session.

Though the music video was the catalyst, the original impetus for this rising desire to materialize his values came from an affecting encounter he had the year before. Saul is a social worker. He assesses the needs of vulnerable adults and arranges care packages for them on behalf of his local authority.[20] This was 2013, and financial cuts to public services were hitting the care system particularly hard.[21] He had already seen his caseload jump from thirty-five to one hundred as staff numbers were being reduced, and he was asked to work longer hours. The encounter did not come through his work, however, but a man he was helping independently through addiction.

> He had no food and his benefits were going to be sanctioned because he didn't turn up for an appointment.[22] I could see that he was being proactive and it just seemed really brutal they were going to stop . . . I remember going out and buying him food and thinking, "I'm paying into this tax system so that people are looked after, and I'm paying twice because you're going to stop this guy's benefits." It started to really occur to me, that's like you are consciously starving somebody. You are *consciously* starving them, and that's our own system that does that. So these kinds

of experiences started to take root, and the sense of injustice was then more than just theoretical. It was a sense that this is actually real. You just have to open your eyes to it.

Saul says he has an unusual instinct for empathy that has been present since childhood, and which has been developed through his exposure to Eastern religious traditions. But he also links his own biography to the predicament of this man. He himself began drinking heavily at the age of sixteen. By the time he was at university he had moved onto harder drugs, recalling with a shudder an anxiety attack that lasted five days long. It was after he graduated that his father decided to intervene. Being a committed member of the Salvation Army, a Christian church with roots in English Methodism, he enrolled his son in its twelve-step treatment program, and it was this that carried Saul through the long process of sobriety. An enduring element of the program involves helping other people through their own addictions, and this was how he met the man being sanctioned.

During his recovery Saul's interest in spirituality bloomed. He started reading key texts of the consciousness movement and learned how to meditate.[23] Although Saul is the third generation of his family to be involved with the Salvation Army, this new direction carried him away from organized Christianity. He started to follow teachings that were rooted in Hinduism, Jainism, Sikhism, and Buddhism—while also developing a new opposition to institutional religion. As Saul articulates his philosophy, it becomes more apparent why he felt so aligned with the ideals of Anonymous, which has substantial overlap with the tenets of New Ageism.[24] He stresses the importance of removing oneself from "identifications" and "attachments," describing social distinctions such as gender as "mental constructs."

Anonymous's own discourses share this repudiation of difference, the original claim to anonymity being a statement of radical sameness, in which all social categories are replaced by a seamless collective consciousness. As with other Anons, Saul's aim is to elevate the level of consciousness among non-Anons as the primary political act; however, for Saul as for AnonUK more broadly, this could result in such an intense value placed on individual action that effective collaboration becomes impossible.

In the wake of his first Million Mask March, Saul parted company with Anonymous. He arrived dressed in his best as the circulating adverts had suggested—wearing a shimmering gray suit, green tie, and leather shoes—but just a handful of others arrived likewise.[25] His plan was to guide another meditation session at the start of the march like the one he had done six months beforehand, but despite his promotion of the event on Facebook, other Anons were unresponsive. Saul is left seated cross-legged on the square by himself, holding a simple sign that reads "keep calm and meditate," his singing bowl nearby, while the carnivalesque protest streams by paying little heed. Afterward, he expresses doubts about the presence of compassion among some of the protesters. He describes a scene outside the Houses of Parliament, where a car was encircled by a tight throng of people. A smartly dressed man and a woman were inside looking distinctly uncomfortable, and those around them started to heckle, "Look at them politicians!" Saul remembers them yelling, without evidence that they were. He keenly sensed the discomfort of the pair and took action, pulling his mask down over his face and ushering people out of the way so that the car could move away. The evening was decisive. "This is where I get off," he says. True to his word Saul disassociated in all practical ways from AnonUK after that, returning his extracurricular time to the meditation rituals that had guided him there in the first instance.

Unemployment

Gregor is a bundle of charisma. It is a bracing February day in central London, and he has joined a protest outside the headquarters of the organization responsible for assessing disabled peoples' right to benefits.[26] The activists are livid. They describe being assessed by people who are not medically qualified, and they speak emotionally of friends and loved ones who have died after being deemed "fit for work."[27] Printed paper photographs of the deceased have been set on the ground at the base of the building, framed in a funereal display of candles and tea lights. Gregor's energy is a bubbling tonic to the scene. Wearing a blazer over a hooded sweatshirt, his Guy Fawkes mask astride a white baseball cap, he darts in and out of the crowd, hopping high up onto ledges and back down again, smartphone outstretched like a tuning fork. On one of these circuits he pauses right beside me and talks affably but lightly about what brought him here, his smartphone still commanding much of his attention. Stretching his arm out to drop a nugget of gum in my palm, he sums it up in an accent laced with cockney, "At the end of the day, you're judged on how you treat people weaker than yourself, and we want to stand with these people here."

Gregor grew up in Greater London in the 1970s and 80s. He was raised in poverty, recalling weeks subsisting only on soup, and his mother (to whom he credits his exuberance) died when he was four. His father was immersed in the far-right movements of the period, and he spent most of Gregor's childhood in prison. As a consequence Gregor lived in a house with his uncles, who along with his father and grandfather were "some of the top Skinheads in the South-East"— not only a part of these far-right movements but central to them.[28] They sought to inculcate him into fascism and taught him how to raise his arm in a Nazi style salute. Gregor left school at sixteen with three GCSEs, one of which was an A in Performing Arts. He had always struggled with written subjects, but acting was something he could easily relate to, and he briefly toured with a theater company on stage. Without sufficient qualifications a thespian career never materialized, and in its place he began a string of manual jobs: working in nightclubs, prisons, and construction.

Throughout his early adulthood Gregor remained committed to the democratic dream of hard work's reward. Increasingly estranged from his family, he sought to transition from his own "working-class" background to be "aspiring middle-class." By 2007 he felt he was making headway. He had been hired by the human resources department of a high-street clothing store to train the company's newcomers. "I was just being a so-called hard worker and believing everything the mainstream told me," he says, "The complete opposite of what I am now."

The break came in late 2011. In November that year the British government introduced a controversial welfare policy, in which people receiving unemployment support were compelled to work for free in extra-governmental organizations to retain access to their benefits.[29] Two women arrived in the store under the auspices of the scheme shortly afterward, and in the course of the training they accused Gregor of discrimination. After almost five years in the role, he was rapidly sacked. This experience carries a heightened emotional charge in Gregor's account, particularly the two-year period of unemployment that it instigated.

> From 2011 to 2013 was the interaction that *changed* me. Having to deal with the Job Centre and the way that they handle people and seeing that there is no work out there. It's low paid, zero-hour contracts.[30] It's agency work in a job centre and I'm like, "That's not a job. You're not offering me training. You're not helping me to up-skill or get back into a workforce." And then I was like, "Well, I've got time now." Because I'm hearing this from the DWP.[31] I'm hearing this lying, and then I'm hearing, "Everything's well," on the television, I'm hearing, "Everything's great," and I'm like, "This isn't right. There's something wrong here."

Studying the processes through which revitalization movements emerge, Anthony Wallace maps out a sequence that can be read both individually and collectively.[32] The transition toward these phenomena, for Wallace, begins with a prolonged experience of stress, arising from a change to which the person or community is unable to adapt. Following this, he argues, comes a disturbing experience of "cultural distortion," during which there is a marked discordance between normal cultural expectations, and what has now become possible.[33] In this phase he says, "The elements (of culture) are not harmoniously related but are mutually inconsistent and interfering. For this reason alone, stress continues to rise."[34] The disharmony between Gregor's primary experience of being out of work, and the way contemporary events are being narrated in British public life, recalls Wallace's sequence. Distortion produces, Wallace argues, a loss of meaning and an apathy toward adaptation that ultimately propel a more radical disillusionment with the system (what he calls the "mazeway") as a gestalt.[35] When he was "hung out to dry," Gregor continues, the desire for social mobility that had propelled him to that point collapsed. "Why am I bothering with this system?" he began to question himself. "I've done everything I could possibly do."

Through Anonymous, Gregor becomes more visible than he may ever would have as a white-collar worker. His facility for performance finds a new outlet online and he starts video blogging. He would "just hit record and start talking," about what he was "learning," what he was "seeing." The Million Mask March was the first protest he had ever been to, and the sense of togetherness overwhelming. "The hive mentality, people singing from the same hymn sheet, moving as a group and a collective understanding that s*** is wrong and we need to change it as quickly as possible." After that he moved from blogging to live-streaming. He has a verbose and self-satirizing style that can transfix his audiences, and his livestream account accrues a several-thousand-strong following. Gregor features in reports about Anonymous produced by mainstream media outlets, as well as low-budget documentaries. Yet in this new visibility he is altered. The left-wing ideals of Anonymous prompted him to repudiate the right-wing influences that shaped his past, and his presence at the disability demonstration is part of a new concern for other victims of state oppression. While the possibility of making a complete political reversal is moot, it was the case that joining Anonymous instigated a significant rupture in Gregor's life that is apparent in numerous ways.

Structural carelessness

Traditional initiation rites manufacture a crisis in the life of the initiand that pries open the possibility for change. By deliberately placing the initiand in a state of symbolic death, through an application of mild to extreme practices around their body, the process of remaking personhood is begun. In a ritual rite, however, these discomfiting or painful experiences are normally succeeded by others of being cared for. In the aftermath of physical mutilations, initiands may be treated with balms or other remedies, and prohibited from engaging in activities likely to cause infection; while the length of their seclusion may correspond to the time it takes for wounds to heal.[36] After the psychological disturbance of invisibility, initiands may re-enter their communities as part of some grand public spectacle. Jean La Fontaine argues that the underlying purpose of traditional initiation is to reproduce existing forms of social and cultural authority.[37] By acting in ways that are uncomfortable upon the body of the ritual subject, and then positioning themselves as the caring antidote to these same bodies, those administering the rites control a crisis in the life of the initiand which then positions them as its resolution, thereby reinforcing themselves and their ideas.

The digital initiations into AnonUK operate through an inverse kind of process. Rather than an individual crisis in the life of the initiand that is socially manufactured, digital initiation becomes possible when a social crisis is individually manifested. Rather than being resolved with care, these digital initiations are set in motion when some form of carelessness is experienced and exposed. Carelessness can theoretically arrive inside any social nexus—for Patrick it is the nexus of grassroots alliances to whom he looked for support ("the hard men of this city"); for Saul the lack of care appears in the state benefits system; and for Gregor, it appears at both the workplace and subsequently at the job center. In these stories and those of others, the nexus can be constituted by what Mark Granovetter calls "weak ties"—stranger sociality through abstract relationships—as well as "strong ties"—the bonds of community, intimacy, and kinship.[38] Both have the potential to be transformative. This said, when considered as an aggregate, there is a gendered tilt. Men are more likely to narrate a breakdown in a nexus of weak ties, while women are more likely to narrate breakdowns in strong ones. However the break came, AnonUK offered the solution, for a time at least, to a social crisis that otherwise remained unresolved. Where, following La Fontaine, it can be argued that structural care creates the conditions for social and cultural reproduction, structural carelessness creates the conditions for social and cultural schism, in this case through the powerful appeal of Anonymous.

While there is no single rule, which is part of what makes joining AnonUK a complex phenomenon to map, the most common denominator in these narratives and others revolves around British government decision-making from 2010 onward. In each of the three above, the narrator includes some direct or mediated encounter with a lack of care exhibited by the British state (although in Patrick's, it was equally this that saved his own life). What these imply is that the various political or religious ideologies to which adherents had historically been exposed—from Patrick's anti-fascism, to Saul's Salvation Army Christianity, to Gregor's immersion in skinhead movements—are less foundational to becoming Anon than were these encounters. Moreover, there is often a marked symmetry between the social crisis that was experienced, and the kind of social remedy subsequently sought. This is most apparent in Gregor's story, where his own encounter with the DWP ultimately led to his presence at a protest organized by disability activists who shared similar experiences of the institution. Ideologically open to new adherents from across the political spectrum, there is not one route of entry into AnonUK, but given the circulating critiques of the British state within it, those with some direct experience of this are carried closer to its primary concerns.

Before moving onto the forms of online experience that come to occupy this space of separation, it is worth recapitulating what these stories suggest about cognitive susceptibility. Saul's narrative contains two important ingredients for joining AnonUK—the first an experienced insight into some form of carelessness, and the second, a set of online experiences, including a rousing video, that propel him toward in-person political action. Manufactured hunger can be an important part of the preparatory stage of initiation. It may be combined with other physical prohibitions such as the denial of liquids, sleep, rest, coitus, or the ingestion of emetics to induce vomiting.[39] In religious initiations, these signal a move away from a mundane world of everyday life, to another populated by gods or spirits. But in the ritual subject these practices can also produce altered states of consciousness, making them disoriented or lightheaded, transforming their own internal thoughts or sensations, and making them more receptive to knowledge. In Saul's account, however, this experience of hunger did not take place in his own body, but in the body of someone else. Rather than his own embodied knowing, Saul is accessing this experience remotely through thinking and imagining. Indeed it appears to be the same empathetic capacity that allows him to switch positions again at the Million Mask March. As he says of the couple in the car, "I put myself in their shoes." And once he was in them, he walked "off." For others, once on, there is far less option of getting off.

DREAM

That life changing moment when you start seeing
The world for what it truly is

Matrix meme (author unknown)

"Searching . . ."

The words appear on the computer screen. Neo is asleep in front of it, his right cheek resting on the desk, as the greenish glow of its moving images falls across his face. Among these is a news story about a character called Morpheus, who shares a name with an ancient Greek deity who visits human beings in their dreams. The screen goes dark.

"Wake up, Neo . . ."

The instruction mysteriously types itself and Neo's eyes blink open, instantly alert. He raises his head slowly and considers the computer more carefully.

"The Matrix has you . . ."

The typing continues. He looks about him, now visibly disturbed, and presses Ctrl X—the cut command—in an effort to restrain the computer's strange new autonomy.

"Follow the white rabbit."

He repeats the phrase below his breath in a gesture which looks like an effort to suck meaning out of it, then presses the Esc key to halt this autonomy completely. Nothing happens.

"Knock, knock, Neo."

Two loud thumps are heard at the door to his apartment. The sudden movement from the virtual to the physical realm jolts his back straight with a start. Tentatively, he opens the door.

In the midst of these crises, Anons describe profound and theoretically formative experiences online, particularly on the audiovisual file-sharing platform

YouTube. Although reading and writing may constitute an element of these experiences, those recalled as the most influential are consistently ocular. What is seen online is able to furnish a fundamental truth, through which their understanding of the world is illuminated and reframed.

Falling "down the rabbit hole" has become a popular metaphor for online immersion, especially on YouTube.[1] Originally a reference to Lewis Carroll's novel *Alice in Wonderland*, in which a young girl accidentally enters a surreal dreamlike world after following a white rabbit down their burrow, the image was repurposed in *The Matrix* to represent a new paraworld of computer-mediated sociality. Here the Wachowskis were following an existing tradition of portraying virtual experience using threshold metaphors, one that extends at least to William Gibson's notion of cyberspace, as a space out of space where different laws apply.[2] The way that Anons describe their experiences online is not commensurate, however, with these casual and playful references to falling down the rabbit hole, where Internet users spend a few lost hours online before resurfacing, perhaps regretfully, but in the final analysis unchanged. In the midst of crises like those documented above, Anons recall spending weeks or even months online, searching for answers to social and cosmological questions that the crisis is provoking. For Anons, the rabbit hole on YouTube and other audiovisual platforms is neither casual nor playful, but presents a meaningful threshold, essential to the transformation they call waking up. Indeed, when those in AnonUK define what Anonymous is, they occasionally draw on threshold imagery. It is a "window," a "front," a "foothold," or a "stepping stone"—objects that exist between one kind of place and another.

Observing the symbolic significance of crossing physical boundaries during rites of passage, Van Gennep develops the concept of threshold rites.[3] Assembling a range of examples, he shows that these physical crossings mark the transition of a ritual subject from a profane world of everyday life, to a sacred one where fundamental changes can occur. The boundaries may be organically afforded by the natural environment, such as mountains or rivers, or they can be symbolically produced, like a bundle of herbs or a stake in the ground. What matters most is that they are considered *as* boundaries by those choreographing the ritual. During the liminal period of initiation, Turner argues, material culture again plays a central role, as sacred objects or *sacra* are shown to the initiand, with the aim of transmitting sociogony and cosmogony. To consider the existence of a *virtual* threshold, in which the only form of material culture is a human body and a computer screen, is to consider a very differently mediated encounter. However, these secondary descriptions suggest that the encounter between subject and screen is able to furnish the digital initiand with theories of the world, in ways that are comparable to this "dreaming" period of the rite.

To theorize these ocular experiences online, I draw on the concept of witnessing.[4] To witness, an idea which dates to antiquity, is to access an immediate truth. As Kelly Oliver and John Durham Peters separately observe however, this overarching idea in fact conceals two very different, if not mutually contradictory, conceptions of how this truth comes to be known. On the one side lies the juridical concept of eyewitness testimony that privileges the knowledge acquired through sensory experience, a concept which feeds into the truth claims of experimentalism during the scientific revolution. On the other, in religious Christianity, bearing witness entails knowing something internally that cannot be seen. When preachers ask for a witness, Peters remarks, "They invite audience affirmation and participation." Here witnessing is not first and foremost a sensory experience but "a public gesture of faith."[5] As an avowedly secular phenomenon, when those in AnonUK describe learning something true about the world by seeing it online, they do so in a discourse indicating that the genealogy they draw is to the sensory truths of law and science. But the actual forms these ocular experiences take, particularly in a willing submission to the authority of online speakers, betray a far closer resemblance to this Christian heritage. The significance of interior, or subjective knowledge, is revisited in chapter 5.

This double meaning comes through in the scene above. "Searching . . ." is ostensibly a reference to the technical procedure of the Internet search. Yet the story of Neo and the transformation he goes through, a process set in motion at his computer, is a narrative of spiritual search. Through it he learns foundational truths about the world that, at an unconscious level, he already knew. Deliberately or not, Neo's transformation in *The Matrix* is shot through with initiatory motifs. After opening the door of his apartment and exiting with a group of ravers, Neo crosses a number of other physical thresholds—a window, a tunnel, a staircase and two more doors, before making his final crossing through the surface of a mirror, waking up naked and hairless like a baby inside a quasi-womb.[6] As with other online cults, *The Matrix* is a frequent reference point in the spaces of AnonUK.[7] Anons identify with Neo and the change he goes through, a change that arises out of his virtual other life. At this, we turn to the story of Kate.

Kate's Awakening

It is a warm June day in a city in eastern England. Kate is sitting on some shallow stone steps, resting her head between her knees in a gesture of self-soothing. When she sees me approach, she clambers slowly to her feet, fighting the fatigue across her face with a gentle smile. She is not getting a lot of sleep, she says. At the moment she is squatting in an ex-council building with a number of other Anons and has been managing on only three or four hours a night. You would not guess, though, at her living arrangements from the way Kate presents herself, as she bears no obvious hallmarks of a counterculture—no tattoos or piercings, no patches or badges, indeed no political signifiers at all, least of all the Guy Fawkes mask that she has worn on other occasions. With her straight shoulder-length brown hair, lilac cotton top, and bootleg blue jeans, Kate blends in easily with the city's other lunchtime consumers. As we park ourselves down in the shady space of a roomy café, its glass walls folded to the sides to embrace its contiguity with the street, there is little hint of the story to come as Kate orders a small cappuccino.

As a child, Kate grew up a few hundred miles away in a seaside town. She was raised by her father and stepmother after her parents divorced when she was six, in what she describes as "a really strict Christian upbringing." Both were evangelicals who worked for their local Salvation Army: her father as the manager and her stepmother as one of the cooks. Naturally this meant that she too participated in the life of the organization, attending Christian activities twice a week and playing the tambourine in the Army choir. As she grew into a teenager, her relationship with her stepmother became more difficult, and eventually her father turned her out of their home, telling her to go and live with her biological mother. This was not a workable solution either, and so by the age of sixteen Kate was living independently. She left the formal school system straight after receiving her GCSEs and was soon working a full-time job to meet the costs of living.[8] Her life continued in much the same pattern until she was twenty-eight, as she shifted between a variety of white-collar roles and was "never political at all."

The year was 2013. Kate was "going through a really bad patch" (she does not elaborate on why) and was out walking in the rain when her feelings overtook her. "I literally physically shouted 'Somebody!'" Suddenly her brown eyes prick with tears and she turns her head briefly away, apologizing for becoming "emotional." She gathers herself swiftly and goes on. "I was like, 'Somebody just help me and guide me to what I'm supposed to be doing.'" It should be remarked that by this point in her life Kate no longer identified with any formal religion, so while the question may have had its roots in her Christian past and its regular interpellation of a deity, now she possessed no obvious interlocutor, spiritual or otherwise. As a result she returned home, her clothes damp and her appeal unanswered.

Two weeks later on the cusp of Christmas, she was up in the small hours of the morning watching YouTube. It was then and there that her "awakening" occurred. Indeed, in literal terms Kate's awakening was just that, as it was the insomnia she periodically suffers from that was keeping her up, nudging her toward the flickering images of the platform to keep herself "entertained." In the midst of these an interview "came up."

> A lady was talking about a document that had been released onto the NASA website called Silent Weapons for Quiet Wars or Quiet Weapons for Silent Wars. I don't know. Something clicked, and I was like, "What

the hell is this about?" It just intrigued me and I started looking into it a little bit more. I just felt that there was something in it that was worth exploring. That was it really. I've never looked back since that, and now I've just become more and more interested in what's actually, really going on in the world.

Although Kate cannot recall the exact title of the video, it was this pivotal clicking moment that introduced her to one of AnonUK's most pervasive doctrines—The Depopulation Agenda.[9] Just less than a year later, Kate bought her Guy Fawkes mask and became involved in the local and national street activities of Anonymous.

Just as Kate is expanding further on some of the day-to-day changes her "awakening" has wrought, she spots a young man wandering through the street and calls him over. He is wearing a gray tracksuit that hangs off his thin frame, a dark patch of sweat tracking the line of his spine, and gray socks with no shoes on. Mike is homeless and has been cohabiting with her in the squat. He seems to be filled with a kind of nervous energy, talking to her briefly before walking away, but then circling back two more times to continue the conversation in snippets. He is keen that she come with him to a forthcoming meeting he has at a sandwich shop, to arrange for the distribution of surplus food to the homeless residents of the city. I do not reflect much on this encounter until later, when on leaving the café Kate spots Mike lying slumped and unconscious in a doorway. We go straight over and although he is breathing, his eyes are half-closed and unseeing. Kate attempts to rouse him by tenderly pinching his earlobe, but without tangible success. She shoots me a worried look. We quickly learn what has happened when she discovers an empty packet of Diazepam and some Mirtazapine pills that had been prescribed that day in his pocket—drugs used to treat anxiety and depression and which can cause drowsiness. After some time a small crowd gathers including a paramedic, and Mike eventually becomes more *compos mentis*. Yet in spite of all the new solicitous presences he remains wholly focused on Kate, whom he appears to trust implicitly. "Mike, do you want to lean on me?" she asks calmly, presenting her left shoulder for this purpose. Once we are satisfied he is being looked after we take our leave. But within just a hundred yards Kate meets another homeless man she knows and descends naturally into conversation, squatting down on the pavement to speak to him and exchanging a two-armed embrace before she rises.

The Virtual Threshold

Kate's account of a decisive audiovisual experience online is the modal route of entry into AnonUK.[10] **Table 1** contains a comprehensive list of all the films that Anons cite explicitly, and her film falls into a wider genre of these—documentary films that present a holistic theory of the world in order to fundamentally reorient

TABLE 1. Complete list of audiovisual media cited

GENRE A HOLISTIC THEORETICAL FILMS	GENRE B POLITICAL DOCUMENTARIES	GENRE C AUDIOVISUAL CULTURE BY/ABOUT ANONYMOUS	GENRE D FICTION
Earthlings (2005)	All Watched Over by Machines of Loving Grace (2011)	Anonymous: A Million Men (2015)	2012 (2009)
Ethos (2011)	Capitalism: A Love Story (2009)	Anonymous makes homeless man cry (2015)	Born on the Fourth of July (1989)
The Fuck-it Point (2012)	Concrete Heart Land (2014)	Anonymous: The Movement Behind the Mask (2014)	Braveheart (1995)
Thrive: What on Earth will it take? (2011)	Enron: The Smartest Guys in the Room (2005)	Omnipotent (2014)	Dexter (2006)
Where are we going? (2009)	Five Broken Cameras (2011)	Anonymous—Operation Onslaught (2011)	Fight Club (1999)
Zeitgeist: The Movie (2007)	Forks Over Knives (2011)	We Are Legion: The Story of the Hacktivists (2012)	Gladiator (2000)
Zeitgeist: Addendum (2008)	The Power of Nightmares: The Rise of the Politics of Fear (2004)		Groundhog Day (1993)
Zeitgeist: Moving Forward (2011)	Power of Nightmares		Pitch Black (2000)
	RIP: A Remix Manifesto (2008)		Return of the Jedi (1983)
	The Century of the Self (2002)		Snowden (2016)
	The Corporation (2003)		Spartacus (1960)
	The Keepers (2017)		The Fifth Estate (2013)
	The Mayfair Set (1999)		The Great Dictator (1940)
	Traffic (2014)		The Lives of Others (2006)
	Unlawful Killing (2011)		The Matrix (1999)
			The Truman Show (1998)
			Tomorrowland: A World Beyond (2015)
			V for Vendetta (2005)
			The X Files (1993–2018)
			Wall Street (1987)

the viewer's relationship to it. These films are listed under Genre A, and I explore one of them in more detail below. Another woman relays a story not dissimilar from Kate's.

> The thing that woke me up was a video called Thrive. Someone sent me a video called Thrive. It's well worth watching. A brilliant wake-up video because it covers everything. Very simple terms but it covers everything—from the wars, money, pharmaceuticals . . . As soon as I watched it straight-away I took it all on board and I could see what they were talking about. And then you just go down the rabbit hole don't you? One thing leads to another.

In most accounts these films are happened upon independently by clicking on machine-generated hypertext—just as Kate's video "came up" on YouTube. But one aspect of this account is not unique: this woman was sent the video by someone she knows, suggesting an interplay of distanced and more immediate relationships.

There are at least two further genres of film that pave a path into AnonUK. The second is the documentary film about a specific political issue—frequently marginal to the mainstream or unknown—with a narrative arc that seeks to produce an affective response in the viewer, and through this to advance its own political project (Genre B). One man talks about how he got his Guy Fawkes mask after watching a film about occupied Palestine.

> It's upsetting to watch but it's important to watch. I took away so much from that and that put me along the lines of "That's why I have one."

Finally into the third genre we can place the large volume of audiovisual culture by or about Anonymous (Genre C). As a loose association of Internet users, Anonymous has itself been a prodigious producer of audiovisual content, most famously in the form of a masked newsreader who issues political statements and warnings. But there are also other forms of audiovisual culture such as music videos and mash-ups which are cited as significant. One of these generated part of the impetus for the first Million Mask March.[11]

> In June 2011 there was a video released by Anonymous . . . It just seemed to capture everyone's imagination and to mobilize thousands of different people across the world. At that point I was literally saying, "I'm going to go to a protest soon," saw that video and thought, "Now's the time."[12]

There have also been a number of documentary films about Anonymous, of which the feature-length film covering the first wave of hacktivism constitutes for many the emic history of the movement.[13] It should be remarked that none of these genres are mutually exclusive, and some testimonies include all three.

Many of the historic forms taken by both religious and secular have been significantly mediated by the printed word, particularly a single book.[14] By contrast, AnonUK is oriented around the digital moving image in all its multiplicity. Anons rarely, if ever, cite print literature in conversation, and if they do, it is often critically, both as a medium and as a source of reliable knowledge. For example,

> I remember going to school with a satchel with that many books in it, cutting into your arms and it was so heavy. But now you can take a phone and get everything you need to know.
>
> I don't believe in books because they are somebody else's opinions, and somebody else has already spoken about them.

Because the digital is the vehicle for truth (an axiom discussed in chapter 5), other media are epistemologically subordinate. But the relationship to digital audiovisual content is not simply about truth claims, it is a significant basis for shared cultural life. Several Anons express an explicit preference for watching over reading—either because they find reading difficult, or they are naturally "visual" people. Alongside the citation of documentary films to make particular factual claims, then, they may also cite fiction films and programs to reinforce conceptual points. A full list of these films can be found under Genre D, most of which contain liberatory narratives of individuals breaking free of their constraints, or social parables about the illusory nature of reality.

The actual mechanisms of transformation during the liminal period have always presented an obstacle to scholars of initiation. In traditional initiations, the frequently secret nature of the knowledge imparted means that presence at these rites may be strictly controlled, having pragmatic implications for access. In these cases, anthropologists must either be initiated themselves, or already be culture members, to be able to document this period ethnographically.[15] Here there is a comparable obstacle for different reasons. While the knowledge Anons access online is nominally public, it is virtually impossible to record these transformative audiovisual experiences first-hand, because they usually take place in intimate spaces, and in an intensely individualized manner whose significance becomes clear only retroactively—sometimes many months after the event.[16] Although I watch a number of films online with my interlocutors, I was not present, either virtually or in person, during any experience described as transformative, but only when these transformations were being re-rehearsed. The result is that I rely largely on testimony, which brings with it the issue faced by testimonial—namely, the reliability of memory. This is nothing new. But it is worth restating because the problem of memory seems to become *particularly* noticeable in relation to online experience. Kate's inability to recall the exact title of the film that had such far-reaching effects on her is symptomatic of a wider fogginess around the totality of films one has seen and who produced them. The contents of table 1 are likely only a narrow slice of these; sometimes the only certainty Anons express is that they watched the films on YouTube.

> A lot of it is videos on YouTube. I couldn't say which ones, but I've listened to a lot of people, and I don't even remember their names or the books or anything, but I've just listened to the information on YouTube.

Connected to this fogginess are tales of immersion, in which participants go on the platform to look for a video and re-surface "800 videos later"—an experience that sounds like entering a hypnotic state.

The affordances of the platform show how this might happen.[17] YouTube is sometimes thought of as an archive, as a receptacle of cultural memory.[18] Yet YouTube has neither the commitment to perpetual storage, nor to systematic retrieval that an institutional archive would. The website is structured to accept any video uploaded as long as it is less than twelve hours long, and as a consequence its data centers hold countless billions of films. But YouTube also continually deletes old content, and despite some institutional efforts to retrieve lost videos, this can be at best a partial endeavor, and by some estimates up to a quarter of all uploaded files have now been lost.[19] Not being subject to curation by archivists, the videos on YouTube are also subject to no central cataloging system, and instead rely on the metadata provided by the person or entity who uploaded it. This has been called a "tagsonomy," a filing system that is oriented around self-indexing or "tagging," rather than a taxonomy of agreed categorical distinctions. It is this tagsonomy that makes new videos visible. When viewers watch a film on YouTube, it is always accompanied by a sidebar of auxiliary films that are semantically connected to it through this metadata, an interface that distinguishes it from watching old-style television, where the viewer is locked into the fixed schedule of a particular channel.[20] The overarching goal of the platform is to hold the attention of the viewer for as long as possible, and it does this by continually offering more content personalized to their browsing habits. There is a significant recent body of work that reflects in depth on how YouTube and other platforms operate to command and sustain attention, in what constitutes a relatively recent economic model—the "attention economy."[21]

The infrastructural logic of these platforms is profit-seeking.[22] In a designed disorientation of the consumer, they are continuous with extant spaces defined by the commodity form, such as the nineteenth-century arcades described by

Walter Benjamin.[23] These arcades were long narrow passageways with shops on either side, the precursor to the modern shopping mall, that offered shelter while limiting egress, dominating the senses of the would-be consumer with illuminated goods. Benjamin thinks of the arcades as threshold spaces, drawing on Van Gennep to describe them as an "intrauterine world," marked by clear entrances and exit points, that sought to redirect spiritual needs once met by religion into the service of a capitalist economy.[24] On YouTube and elsewhere, Benjamin's "dreamworld of mass culture" can be refigured as a dreamworld of mass communication that shares the same modality of retention, disorientation, and ultimately commodity exchange.[25] What has been less well anticipated and presupposed however, and distinguishes these virtual worlds from their spatial precursors, are the theoretically transformative possibilities of threshold experience. Although the infrastructural aim may nominally be some form of profit, the effects of these dreamworlds on their audiences may be something else entirely.[26] Benjamin saw the arcades as a substitute for religious experience. Yet the forms of theoretical instruction that circulate on audiovisual platforms render these virtual capitalist domains far more capable of producing experiences that can once again take quasi-religious form. For those like Kate, it is where they go to find answers to the deepest questions of existence.

Ocularity in Comparison

The liminal period of initiation is often considered to confer the ritual subject with powers of sight.[27] Eyes, whether those of the initiand, the initiated, or a third-party such as an animal, are employed as key props to mark the epistemic changes that the ritual has produced.[28] Moreover, the masks that the initiated sometimes earn the right to wear may possess exaggerated ocular features to represent these changes. One example of these is the "Ejumba" mask used in Jola initiations in the Casamance region of West Africa.[29] The Ejumba are worn by newly initiated young men as they stumble out of the forest at the end of their seclusion. The mask consists of a wide-domed headpiece woven from palm leaves, with thin strips of raffia attached to it, that hang down the length of their bodies. On either side of the headpiece are a pair of cattle horns (showing that the initiand has taken on the properties of the sacred bull), and just beneath the forehead are a pair of protuberant hollow cylinders made of plant fiber that look like old driving goggles. These ocular cylinders represent the "acuity of vision" the men have acquired in the course of their seclusion; as for the Jola, becoming initiated means developing the capacity to "see."[30] More specifically, it means being able to see the invisible and potentially malevolent beings of the Jola night world, from whom it is now their obligation to protect the rest of the community.

In his seminal essay, Turner goes deeper into the ritual processes through which these powers are won. During the learning phase of the liminal period, initiates may be shown specific sacred and often secret objects by their elders known as the *sacra*.[31] *Sacra* can take a large variety of forms. They may be relics like figurines, effigies, icons, emblems. They may be musical instruments like drums, or tambourines. They may also consist of objects without any workmanship at all, such as a piece of bone or a sheep's fleece. Turner resolves this wide variety with the argument that what is most important about the *sacra* is not their outward form, but the "instructions" that elders give about them.[32] In other words the substance of their meaning comes not from their material characteristics, but from the narratives they stand for and into which they fit. In religious contexts this would mean stories about the beginning of the universe and the names of gods or spirits, while in secular contexts this would mean accounts of the foundation and emergence of the society into which they are being initiated. Turner draws out the fact that these objects frequently possess some disproportionate feature, such as an oversized head, or a vast pregnant belly. This exaggerated feature presents a visual focus for the lesson in question, that is transmitted with its help. Turner is unequivocal about the importance of these aural and visual experiences in the remaking of the initiand. The acts of communication that take place through the *sacra* are, for Turner, "the heart of the liminal matter."[33]

Many of the documentaries cited by Anons can be illuminated through Turner's analysis of the *sacra*. This is particularly true for Genre A—films that offer a holistic theory of the world. As an online cult oriented around the moving image, Anons constantly recirculate favored films on social media. But there is at least one film on behalf of which they take the more labor-intensive step of recording onto a DVD to disseminate it in person (see **fig. 1**). The DVD contains an unofficial recording of a documentary called *Ethos* released in 2011, which for this reason is worth examining at length.[34]

The opening scene of *Ethos* is a view of the Earth from space as it rotates slowly on its axis. It is a vista which signals the cosmological ambition of the narrative that follows (namely a partial account of the history and structure of American capitalism) and echoes the initiatory emphasis on cosmogony, stories about the origins of the universe. The documentary is organized as a visual essay with clearly identified topics—"Politics," "Corporations," "The Media," and so on—in an attempt to synthesize these various areas of social life within its whole. Significantly, the argument is driven forward at every stage by narrators. The first of these is the actor Woody Harrelson, who functions as an anchor for the overarching story, while there is also another lilting disembodied voice who narrates some of the film's darker moments. In addition to these two recurring voices, the film moves between a number of other talking heads, experts in a relevant field

FIGURE 1. Compact disc containing the film *Ethos*

whether through academic study or insider expertise, who provide its principal substance. These narrations, both visible and invisible, are then combined with moving images which function as animated illustrations—films of dollar bills being printed accompany discussions of money, words highlighted on a page indicate a particular book, et cetera. The source and content of the footage is never cited, so the viewer is never entirely clear what it is they are looking at and consequently relies on the film's sequence of narrators to understand its theoretical implications.

The particular modality through which knowledge is transmitted in *Ethos*—which is by no means unique—deploys a hierarchy of word over image. As with communication of the *sacra*, "what is shown" is ultimately subordinate to "what is said," which is where its full meaning derives.[35] Without any of the citation practices of text, films such as these do not grant the viewer any form of verification for the statements made. In fact they do not appear to require verification (as academically understood) in order to exert their effects. Instead, unnamed archival footage is assembled to illustrate theories of the world that hinge on the authority of narration. What makes this form of knowledge transfer even more contradictory, however, is that while operating through hierarchical instruction analogous to communications by the already initiated in the liminal period, they draw instead on the legacy of experimental science to assert their own truth claims.

Digital Witnessing

The invention of experimentalism as a way of producing scientific knowledge depended, as Steven Shapin and Simon Schaffer show, on a variety of different technologies that were at once material and social.[36] Central to this was a privileging of the senses, particularly the visual sense, in asserting and disseminating truth claims about the behavior of natural phenomena under investigation. Nature could be collectively witnessed under highly specific conditions, and through this act of witnessing, scientific facts established. A substantial precedent for visual truth claims already existed in the law. Eyewitness testimony had long been an active juridical concept, and drawn on explicitly as an analogy to scientific witnessing when the method was in its infancy.[37] What was particular to the scientific method was that what was witnessed was irreducibly collective. Scientific facts became true because a select community of scientists agreed on what they had witnessed together. Experimentation was invented in opposition to the solitude and secrecy of the alchemist's closet on the one hand, and the individual dictates of the philosopher on the other. But what was even more important than these immediate acts of witnessing in the laboratories, they argue, was "virtual witnessing"—the reproduction of scientific images, and the development of a scientific way of writing, that could then multiply the number of witnesses through print almost indefinitely.[38] Through these distant witnesses, whose assent was based on trust, the truth-claims of science were expanded.

The enduring potency of eye-witnessing, in science and the law, to assert truth claims, creates space in which this can be claimed by political actors. Witnessing, Kelly Oliver offers, is a form of subjectivation that can work to repair the traumas arising from a lack of political recognition.[39] In her ethnography of animal rights activism in India, Naisargi Dave recounts stories, not dissimilar from those of Anons, in which activists describe transformative ocular experiences—in this case immediate experiences of animal suffering—that mark decisive turning points in their political biographies.[40] Indeed, in Dave's analysis and in her interlocutors' accounts, there are several initiatory motifs. "To witness," she explains, "might best be understood as a radical interpenetration of life and death."[41] The witnessing of the biological death of a suffering animal cleaves a rupture in the life of the activist, akin to the symbolic death of their former self who departs along with it. In one of her examples, this moment of symbolic death is combined with scarification, when an activist drops a spoon of scalding hot soup on her wrist, leaving a burn marking the exact moment of her epiphany. Witnessing is distinct from voyeurism or spectacle, she argues, because it implicates the witness in the morality of what they have seen, leaving these activists morally transformed.

A debate exists among scholars of witnessing, around whether media witnessing can be as potent as these more immediate forms. Alissa Richardson, in her elaboration of what she names "black witnessing"—the viewing of audiovisual recordings of racialized violence among Black people—argues that these are equally important vectors for politicization.[42] Her argument is charged with extra force by the development of the Black Lives Matter movement in 2020, and the various individuations of Anonymous also make this case. Indeed, just as AnonUK was contracting, another Anonymous was being created by two Australian animal rights activists to increase the number of vegans across the world, founded upon this premise. In the self-named "Anonymous Voices for the Voiceless", or AV for short (an abbreviation that similarly stands for audiovisual), digital witnessing became a concrete political technology.[43] AV choreographed a performance that could be replicated in any public space, in which activists wearing Guy Fawkes masks stand in an outward-facing square, displaying images of animal suffering. While some of these may be printed on paper, it became increasingly common for these activists to hold digital screens, on tablets or laptops, showing rolling slaughterhouse footage. People who paused to watch could then be approached by mobile outreach teams, to pass on more information about veganism. This different genre of Guy Fawkes activism was built around the presupposition that what was seen online could produce meaningful political change in the life of the viewer that drew equally strongly on tropes of truth.

The nonfiction films cited in table 1 achieve an epistemological effect by inhabiting some of the aesthetics of the scientific method. They include forms of exposition construed as public, and as collective, and in which virtual witnessing through relations of trust has no lesser claim on the truth. But beneath this aesthetic lies a different idea of what it means to see something, in which viewers are also construed as witnesses to that which cannot be seen, knowledge that is somehow secret and suppressed. Rather than asserting continuities with scientific and legal epistemology, it becomes vital to unpack the distinctions. The experiences online recalled by Kate and other Anons express, in Shapin and Schaffer's terms, very different forms of social organization from those of experimental science, bringing these knowledge practices closer again to individual philosophy, and the seclusion of the alchemist's closet. Moreover, rather than direct eyewitness testimony, these forms of witnessing are heavily curated, a potential much more readily available in digital witnessing. This takes place first and most clearly by the filmmakers themselves, but less obviously though no less influentially, by audiovisual platforms like YouTube. The viewer is placed in an epistemically subordinate positions to these curators, while this asymmetry is effaced through appeals to ocular autonomy.

Anons' encounters inside the virtual threshold recall a dream-like experience from which they eventually "awake." Like Neo falling asleep at his computer, the virtual offers an arena in which fundamental theoretical change becomes possible. The experience of becoming lost in streams of hyperlinks, and then emerging with the conviction that what they have seen constitutes a deeper truth, recalls the way that dreams have been analyzed in the history of psychoanalysis, as a nonlinear symbolic experience that yields profound insights.[44] But dreaming does not need to be framed in this Eurocentric way. References to dreaming are pervasive in initiation.[45] This is especially marked in the societies of Australia, where initiands are said to enter The Dreaming, a separate spatiotemporal realm at the beginning of the universe that they temporarily inhabit in order to comprehend the role of divine beings in creating it. The hyperindividualized quality that dreaming takes inside the virtual threshold finds its closest comparators in the dreams and visions of those entering spiritual professions such as shamanism, where incumbents have autonomous self-reported encounters with gods or spirits, offering them instruction.[46] Kate's "lady," who appears in this moment of perfect solitude in the middle of the night, although algorithmically generated, arrives like the vision of such a being—before melting back into the oblivion of YouTube's ocean of uncataloged content.

For Kate, the British public sphere acquires a very different meaning from the one it had beforehand. Indeed, around the time of these utterances the ghost of Kate's former self is still active online, in the form of a professional profile page she once used to apply for white-collar work—permanently projecting her "passion for products to sell, sell, sell." The timbre of an energetic salesperson bears scant resemblance to this slow-moving woman, now living in a squat and articulate in her criticism of the "money-driven system." On this day at least, the high-street shops may as well be invisible. The city center is not a place for mass consumption, but a place where homeless people live, people whom she actively cares for and about. Kate's dreams have, for the time being, diverged significantly from the modal desires of mass culture. "My dream is to go to Peru and drink Ayahuasca with shamans," she says hopefully, alongside other more homegrown ambitions such as setting up a community hostel and growing her own food. In the end, Kate's biography takes a different direction from the one she imagines this warm June day, veering toward animal rights activism and away from Anonymous. But what remains constant is that this apparition in the middle of the night rendered a sea-change in her biography which was not later reversed. It is now the moment to look more closely at some of the social and cosmological ideas that Anons absorb inside these virtual worlds.

SOCIETY

A group in the nascent state seeks the truth, and there is only one truth.

Alberoni (1984, 71)

Stand up for your rights . . . Delete the Elite . . . We are not slaves

Protest placard on display at the Million Mask March (5 November 2016)

A figure in a black hooded sweatshirt and a mask sits at a desk with a sheaf of paper. Facing a virtual audience, with stiff gestures and a computer-generated voice, they present an alternative news broadcast.[1]

"Greetings world, we are Anonymous.

With all the latest leaks coming out about the power of GCHQ, the NSA, and Five Eyes it is high time that we showed our faces and said enough is enough.[2]

Between 29th August and 1st September 2014, Anonymous will be holding a mass protest outside of GCHQ in Cheltenham, England, to continue their campaign against the mass public surveillance employed by many of the world's governments, including the UK's.

The UK has one of the world's most intrusive surveillance programs. Quite simply, you cannot go anywhere or do anything without the eyes of the government being upon you.

While we are told this measure worthy of Orwell is for our own protection, we feel that this is simply another lie spun to us, and that this massive invasion of privacy is nothing but a method of gathering intelligence to allow greater control of the world's civil population. The tyranny must end. GCHQ–1984 is not an instruction manual."

In the liminal period, the initiand learns the rules that govern the universe. In explicitly religious rites they are taught the pantheon of gods and spirits who control social and environmental forces. In secular rites they learn about mythic founding figures, and occluded sources of social and political power. They also learn of their own position relative to these beings, and their subsequent role as initiated persons in producing a moral order. This position varies according to the initiating cosmology. In the virtual threshold, Anons discover the existence of a global project known as Agenda 21. Agenda 21 is a malevolent plan either to depopulate the world to five hundred million people, or to "enslave" them, undertaken by a shadowy cabal of miscellaneous actors sometimes called the New World Order. In learning about its existence, they also come to understand their role in this great battle between good and evil, and hence the salvation of the world. A substantial amount of scholarship exists about this theory, which forms the backdrop to this chapter.[3] Rather than evaluating its legitimacy, here I describe some of the pathways through which it arrives into AnonUK, and some of the social and political consequences for those involved.

Harry West and Todd Sanders describe theories like these as "occult cosmologies."[4] An occult cosmology is a theory of the world and its origins that privileges the role of "secret, mysterious, and/or unseen powers" in constituting it.[5] Exploding the distinction between west and non-west, West and Sanders do important work in placing Agenda 21 and its analogues alongside other belief systems globally. In the process they submit that what distinguishes these from cosmologies proper, is that they are partial rather than holistic forms of explanation. Inside AnonUK, these theories do not however appear to be partial. Like other initiands, they offer their bearers a complete and more or less coherent architecture of causation. While there may be quibbles about the details, Anons' overarching agreement on its existence, what Alberoni calls their "unanimity," is a binding and mobilizing force.[6] The main difference between this particular cosmology and those commonly found in traditional initiations is the degree of benevolence. The gods and spirits documented by anthropologists of initiation may be fearsome, but they generally remain life-giving and affirming, rather than posing an existential threat that must somehow be challenged.

Anons' willingness to absorb Agenda 21 can be comprehended through multivalent logics of reduction. While the dizzying multiplicity of knowledge available online, as well as the personalization of its delivery, create conditions for extraordinary epistemic individuation, power is still reified here into a singular demonic adversary. By reducing the complexity of political action into a monist thing-like form, it simultaneously allows for the reduction of intention—what David Robertson in his ethnography of comparable theorists calls their "theodicy of the dispossessed."[7] This theodicy, or logic of injustice, gathers up the chaos and complexity of the experiences chronicled in previous chapters, and reconfigures it as order and intent. This intent reduces them to nothing. As objects of Agenda 21, they are either already dead, or "enslaved." While a strange and problematic word for citizens of a twenty-first-century late industrial polity to choose, especially when their involvement in AnonUK has arisen out of many layers of political autonomy, it assembles more sense if read symbolically. In Orlando Patterson's analysis of slavery, itself influenced by Turner, Van Gennep, and other anthropologists, the phenomenon equates to a form of symbolic nonexistence, that is, an obliteration of value that he calls social death.[8]

It has been argued in cult studies that the conceptual apparatuses these formations offer are somehow compensatory.[9] They endow meaning to forms of suffering that may otherwise be imponderable. In AnonUK, it should be further emphasized, this compensation particularly takes the form of status elevation. Because, while as passive objects of Agenda 21 they may be reduced to nothing, as "awakened" political actors they become instead, as Robertson puts it, a "counter-elite."[10] Falling at the optimum location of both knowledge and morality as visualized in **diagram 3**, they are the only segment of their own tripartite understanding of society who bear the two principal values. This sets them apart from their reified adversary that is the embodiment of evil, as well as the audience of their activities who, while containing the potential for goodness, remain nescient. An initiated person is generally in some way an elevated person who has traveled from one status category to another that is higher or more central. In digital initiation this dynamic persists. Though the discovery of this hidden agenda may at times be frightening for Anons, it is, on the whole, energizing and empowering. Because the stakes are so high, it becomes imperative that they take action now.

This chapter traces one of these endeavors, a protest against mass surveillance called by Anonymous in 2014, which takes place outside GCHQ in Cheltenham. Following the build-up to the protest on its Facebook event page beforehand, it documents the presence of three segmenting pronouns—We, They, and You; who falls into them, and how they are represented and perceived. Through the medium of this event as it unfolds, it maps the kinds of knowledge-sharing conversations that Anons have, with particular reference to one man at the protest called Pete. Finally it documents the presence of humor. An important point of continuity with the tricksters introduced by Coleman, humor distinguishes AnonUK from expressly violent online cults. As a form of play it ultimately defuses the confrontation between them and their great adversary, tilting their political efforts, as chronicled in part 2, toward the peaceful project of awakening.

Three Pronouns

Anons share a discourse of the universal human. They commonly use universalizing terms such as "human being," "human rights," "humanitarian," "human liberties," "humanity," and even sometimes "species" when communicating their political aims and sentiments. Alongside this universalism, Anons also dismiss conventional forms of social segmentation such as ethnicity, religion, nation, gender, sexuality, age, and disability as being insignificant. Anyone can be Anonymous is the unifying claim. In February 2014, a public event page is created on Facebook, to further promote the protest outside GCHQ.[11] The page provides a public platform for extensive discussions around the event for the six months prior to its occurrence. By examining these discussions in their entirety, an alternative to conventional forms of segmentation becomes apparent, as identity becomes subsumed within a simpler triangular world.[12]

Those planning to participate in the protest employ the pronoun We to refer to themselves and one other. But who exactly is this We? The first collective it implies is Anonymous itself and is coextensive with the usage of this pronoun in the five-line statement with which they have long been associated.[13] As the months roll by, however, this collective acquires a clear national orientation, as commentators align themselves with "Britain," "civil society," and even "the taxpayer." This orientation also becomes clearer in a subsequent Anonymous broadcast posted on the page, in which the masked speaker refers variously to "innocent citizens," to "the people of this country," and to "the people of the United Kingdom" as subjects of surveillance.[14] In this characterization, the We are represented as moral members of a citizenry whose rights have been violated, pushing them beyond acceptable limits. Consistent with this, the plan for the protest is meticulously law-abiding. One of the page's administrators posts a monochrome map of the intelligence headquarters, with an uneven dotted line around its perimeter, and a large red arrow with the words "WE ARE HERE," pointing to an area beyond it where the protest will be held. Citing the relevant law, he assures readers that they should be able to attend without hindrance, as they will be located in the "public land outside the dots." What also becomes clear in these discussions is the unequivocal humanity of the We, through repeated use of the word "people" and fond references to one another as "brothers and sisters."

The raison d'être of the protest as it emerges does not fall along conventional lines. There is very little mention of the media (whose presence can sometimes be solicited by activists to amplify their cause), nor does it seem to be explicitly intended for the benefit of British lawmakers. Rather, the event takes on the quality of an encounter, a meeting, a show of reciprocity between two inextricably connected parties. If one half of this encounter assembles itself beneath the pronoun We, then the other half can be collected beneath the pronoun "They." But who are They?

In the first instance They are GCHQ. The event, ostensibly, is a demonstration against GCHQ's mass surveillance program as revealed by American whistleblower Edward Snowden.[15] The Snowden revelations showed that there are likely to be traces of their digital selves inside the Cheltenham building, and they seek in a modest way to redress this imbalance. However, in the build-up to the event, the pronoun seems to possess a concatenating logic, as other entities and individuals are swept into its orbit. GCHQ is linked to "the NSA," which is further extended to "UK and US intelligence," or indeed "GCHQFBIMI5CIA homeland security."[16] Connections extend beyond the security services. They are equally "David Cameron," "Boris Johnson," "parliament," and "corporate governments." In the economic sphere the links are particularly numerous. They are "the banks," "the Bilderbergers," "Elite Powerbrokers," and "the Rothschilds."[17] As the months proceed the list continues to grow, bringing in other extra-governmental institutions such as "the Vatican," "Buckingham Palace," and "the BBC." If the We is tacitly characterized as a homogeneous assembly of law-abiding Britons, They emerges as its structural antithesis—a multiple, but ultimately united, transnational entity that operates beyond the law. In the final analysis all the above actors are considered part of "one corrupt corporate shadow government": also known as "the Illuminati," "the Freemasons," "the Elite," or "the New World Order."

Alberoni identifies the We/They distinction as characteristic of collectives in the nascent state.[18] In these enthusiastic early periods of political and religious life, before structural distinctions reassert themselves, this We is formed through total agreement on a particular idea, what he calls their "unanimity."[19] Unanimity does not necessarily entail homogeneity, indeed this collective We will willingly recognize the presence of differences among them, but some form of overarching agreement is essential to the formation of its collective being. This is inextricably connected, he argues, to the construction of an adversary, "The setting up of a wall of total alienation between 'us' and 'them.'"[20] In Robertson's study of theorists in the UK and United States who similarly commit to the concept of a New World Order, he considers this a form of Othering, in which the Other is constructed as "literally alien."[21] What is worth rehearsing in this particular context is the immense heterogeneity of the Other. For just as the We crosses all historical boundaries of identity construction, They do as well. Unlike other present and past political contexts, where the Other may fall along ethnic, religious, or similar lines, there is no standard identifier through which They can be recognized. Indeed, this concatenating and hidden logic means you never know *who* may be a member of this adversarial group, and who is "one of us."

It is significant that the collective names for their adversaries are regularly capitalized. It indexes the ways in which a complex field of political action, when assembled into a singular enemy, becomes reified—a move with consequences for their humanity.[22] While the We is synonymous with the human, through the language of people and fictive siblingship, They are represented very differently. In the midst of the discussions, some commentators post edited images of the ring-shaped GCHQ building, with other images superimposed on its central hollow to propose extramural connections: for instance, the Freemasonic all-seeing eye, or the circular emblem of the NSA. The page also links to a video reportedly made by one of the administrators, in which the viewer looks down on the building from above while a rumbling explosion rips through it, producing a black and fiery cumulus that slowly engulfs the building in its entirety. Presuming the building is occupied in this imagined audiovisual event, its employees become targets for legitimate violence. The salient point is this. Through the symbolic work of rendering their adversaries thing-like, transforming them into faceless collectives, or presenting them as buildings to be destroyed, the effect is thoroughly *de*humanizing. While the We are human, They are represented without human attributes. This can yield a cognitive dissonance when contravened. For example, the same administrator expresses his "shock" that the BBC have been in contact with him about the event. Such a direct exchange with an organization considered part of this broader group upsets the categorical boundaries between We/They, human/inhuman, that Anons construct.

As the discussions continue, another pronoun moves into view—You. You is the imagined audience of the We whom they seek to enlist against the They. While We and They exist in a relation of binary opposition between humanity and inhumanity, You occupy an ambiguous location between these two categories. The word You is sometimes substituted with two other terms—"sleeple" and "sheeple." These linguistic syntheses convey this ambiguity; as while both are amalgams of "people," indicating the group's underlying or potential humanity, they are combined with two other words, "sleep" and "sheep," to suggest that this is not being fully realized. The You are either unconscious or they are partly animal, specifically, an animal that in a Christocentric context connotes uncritical conformity to the wishes of authority. This ambiguous location of the You is attended by ambivalent feelings toward them, feelings that are simultaneously plaintive and hostile. "What is it going to take to wake you up?" one earnestly inquires. "You are hopelessly enslaved," another warns with more edge. When a member of this audience casually steps into one of the threads, telling the Anons that the security services are there to protect them, he is quickly rounded on. "Go back to sheep zzzZZZZZZ," one admonishes, invoking both unconsciousness and animality at once. This riles the outsider, who persists with greater fervor. The Anons' attitude escalates similarly, but among some maintains the tension that this combative constituent of the You could yet become a member of the We—if only he were not "still sleeping" and "walking around with his eyes closed." Such a critique is not an attack on his essential nature—he remains human—but on the way of life he elects. "Go eat a McDonald's. Buy a newspaper. And enjoy your subservient lifestyle," the interlocutor concludes, words which mark his effective banishment from the thread. Ultimately, the protest is presented as a confrontation between We and They, and a chance, momentarily, to reverse the asymmetry between them.

Outside GCHQ

The build-up to the protest is long, and full of such drama. Elsewhere online, on the UK Anonymous events website, a digital clock counts down to the start time in days, hours, minutes, and seconds for the best part of a year. By the time it comes around, more than six hundred people have registered as attending on Facebook. Yet when, at the designated hour, I reach the bleak tarmac landscape and tall wire fences surrounding the mouth of GCHQ, there is not a solitary mask in sight. This is not to say that there is no one here. Several journalists stand around slightly at a loss, two of whom (from the local newspaper and BBC respectively) canvass me for an interview in the absence of any activists. As I am equally bereft of interlocutors, I deposit myself in a nearby park to take notes.

Not long afterward, the waspy buzz of a helicopter above urges me back to the event. By this point a handful of Anons have now gathered in the area, and I approach two middle-aged men who appear to be engaged in deep discourse. They are facing each other, opposite elbows resting on the waist-high railing beside them, completing the scene's incarcerating feel. The men gesture for me to wait a moment while they finish, before turning their heads slowly in my direction, to rest four watery eyes on me from beneath their white and black plastic masks. The gambit is impersonal. "There is no past and no future, there is only the present," the man on the left pronounces in a soft Welsh accent. His friend continues the thought without conjunction. "Time is an instrument used to control people," he says simply. The conversation continues in the same atypical fashion, flitting seamlessly between facial recognition software, the CIA, the cashless society, and the "families" and "bloodlines" "at the top" of it all. Their calm delivery belies the urgency of its content. "It's coming to an end . . . People have got to wake up . . . The mass of people don't even know." The openness of the trialogue seems to provide something of a palliative, however. As it draws to a natural close, one of the men expresses a sense of awe that we are able to stand there and share such a "marvelous conversation." "We need more people to do that," he concludes.

As the day unfurls, the number of Anons creeps up without ever rising above twenty. The demographic includes both men and women and extends in age from young teenagers up to adults in their fifties and perhaps sixties. Despite this age and gender diversity, there is an ethnic and linguistic homogeneity, as there are no people of color, nor is anyone speaking in a language other than English. In contrast to the slick memes and videos that had been circulated online, the material culture of the protest is mostly handmade, with an immediate, DIY quality (see **fig. 2**). Messages are sprayed and scrawled onto paper and scavenged cardboard. Someone has painted "NEW WORLD DISORDER" onto a shred of white sheeting. After being used as a prop for some photographs, it is eventually staked

FIGURE 2. Photograph of author wearing CCTV headpiece at privacy rights protest (photo by author, anonymized by Benjamin Elwin)

to a thin tree, where it raps in the wind beside the Anonymous flag. Virtually everyone wears a Guy Fawkes mask. Despite these efforts at presence, Friday's protest is characterized largely by its absences. Over and above the absence of hundreds of promised attendees, there is also an absence of singing, of chanting, of any music at all beyond the tinny peals of a smartphone, and indeed of any coordinated political activity among those present—making the small gathering feel somewhat denuded and exposed. The Welshman's wish is granted though, as what is alive are the conversations. A diversity of knowledge and points of view meet on a ground of unanimity about what they are there to fight against. Of these, one of the most animated discussions is a blonde man in a yellow jacket called Pete.

Pete tells his story of separation. He was driving in the early hours of one night in 2008, when he noticed a police car was following him and eventually pulled him over. The officer charged him with speeding, which he asserts was untrue. When Pete queried the claim, the former allegedly replied that it was his word against theirs and pressed on with the charges. Pete was left with a sense of his own vulnerability relative to this inequality in the value of testimony. Since that point he has become concerned with documenting abuses of police authority, and he has been to court several times. More recently, he recalls, he was threatened by police with a taser and spent the night in a cell. The handcuffs he was compelled to wear bound his wrists so tightly that they damaged them, crushing the radial nerves, he says, displaying thin white scars around the area. Pete had originally trained as a scientist, and he now uses his knowledge to invent technologies to protect himself. Indeed, the coat he is currently wearing is one of his own inventions—a taser-proof, portable Faraday Cage—lined with a thin metal that gives it the capacity to block radio waves. He offers to prove it, taking my mobile phone and placing it in his top pocket, after which it duly fails to ring when summoned.

What is characteristic in Pete's account is the rapid slide from defining personal experiences to larger theoretical claims. He spends some time deconstructing the symbology of the British birth certificate, and the legal rituals surrounding it.

> Our birth certificates prove that we are slaves. When you go to court they have to make sure you say your name—no name no game—because we're corporatized. That's why your name's in capitals, back to Roman times. When your name is in full capitals it gives you the standing of an ant. You have a hierarchy, a top man, like the Lord of the Manor, a sort of middle, and then you're just a slave at the bottom. And slaves have their names in full capitals, whereas kings and queens are all in lower case.

Our birth certificate, it says on it, "Property of Crown Corporation." The crown corporation is the square mile in London, all the banks and the people running the show. They want you to admit that you are this name on your birth certificate.

Shortly after, the sense of enslavement he communicates shifts toward another, far greater existential threat faced by humanity as a whole. Talk turns to the UK's rising national debt, and he continues,

If you extrapolate, as any sensible person would do, where's the point where things absolutely grind to a halt? It can only be a few years now . . . If I hadn't seen the Georgia Guidestones, I would have thought, "Argh they can't, they're not going to do that." But the fact they've said they need to maintain the population of the human race at under five hundred million, it's pretty much saying, "Look, at some point we are going to wipe you all out."

People seem to think that they need us. They don't need us. They've got all the resources and manpower they need underground. They'll stay down there while the economy is wiped out, we all fight amongst ourselves, the military take over, and then when there's not much left, because everyone's dead from either starvation or illness, they will rise up and inherit the earth.

These two sets of ideas circulate widely in Anonymous settings. The first is the notion of enslavement from birth, one that is initiated in the first act of legal recognition—the birth certificate—and then continues throughout life through the legal personhood that this certificate confers. The second is what is known as The Depopulation Agenda, a theory that asserts the existence of a plan to reduce the earth's population to a maximum of five hundred million inhabitants. Logically these ideas contradict one another, and there is occasionally dispute between Anons over whether their biological existence is required to perform coerced labor, or whether it is simply disposable. Even Pete's account betrays these contradictions. He argues first that he is an object of property, which would suggest that, as with other forms of property, he possesses something of value to the proprietor. He then argues that his existence is likely to be terminated due to its disutility. One of the things that remains constant between the two proposals is the notion of a grand dehumanizing project undertaken by the They, whatever their specific intentions. It is now the moment to track these ideas more thoroughly and establish their implications.

Vernacular Theorizing

The first set of notions elaborated by Pete are associated with the "Freeman/men," or the "Freeman/men on the Land" movement. Broadly stated, Freemen assert a strict moral opposition between common law, and civil or statute law, maintaining that only the former holds legitimate jurisdiction over them, and the latter can be avoided using legal methods.[23] Having so far eluded extensive scholarly attention, the precise origins of this movement are unclear, but several accounts link it to a specific Canadian man in the early 2000s, from which it spread to other common law jurisdictions: the United States, the UK, New Zealand, and Australia.[24] Stephen Kent traces its key tenets to the far-right sovereign citizen movements that emerged in the American midwest in the 1960s and gathered pace during the farm crisis and then the interest rate crisis of the 1970s and 80s, which led to hundreds of thousands of farm closures across North America.[25] Often with overtly racist strands, these movements oriented themselves around the importance of independence from government and were linked to tax boycotts as well as actual violence against state officials.[26] The political picture is murky, however, because as Kent acknowledges, other aspects of Freeman thinking can be found in anticolonial and peasant movements and their assertion of the right to land. This odd blurring of left and right persists into its British articulation, as one of the first major media appearances of Freemen in Britain is an article in *The Guardian*, penned by an activist at Occupy London.[27]

As an online cult oriented around the veracity of the moving image, Anons' most important encounter with Freeman orthodoxy arrives through a widely watched film on YouTube.[28] It is a recording of an hour-long lecture by a Freeman called John Harris. He is speaking at a conference held in southern England in 2009, in a large public hall to a several-hundred-strong audience. Titled "It's an illusion," Harris's talk attempts to produce an ontological shift in the minds of his audience—constituted ostensibly in the hall, but with much more lasting significance through the film. The shift turns on the essential proposal that what the audience thinks are Britain's democratic institutions (parliament, political parties, the police, the courts, etc.) are in fact corporations. In other words, the primary function of these institutions is not to represent and serve the rights and interests of British citizens, but instead to profit from them for the benefit of shareholders. The illusion is achieved, Harris maintains (in what is a cornerstone of Freeman thought) through the concept of the legal "person." Rather than being the formal means through which Britons access the rights of citizenship, legal personhood is reimagined as the means through which they are owned and commodified by this polity. This is where the birth certificate comes in. Instead of conferring rights of birth, this document becomes the principal mechanism through which the named individual is created as a "bonded slave," and the moment when their parents cede all rights over the child to the state *qua* corporation.

People who identify with Anonymous, and/or attend Anonymous events, explain iterations and extensions of these ideas on a number of different occasions. Although the Freeman movement is known to be somewhat distinct from Anonymous, with its own practices and priorities, its orthodoxy is clearly compatible with Anonymous and moves freely within it (see for example **fig. 3**). Anons variously say that the birth certificate is a government bond, that you can type its number into a website and see it being traded on the stock market, or even that the black ink used in its inscriptions indicates that the bearer is, in actual fact, "dead."[29] The specificity of the deconstruction is slightly different with each retelling—such as Pete's description of a three-tier class system—but where the compatibility lies with other discourses circulating inside Anonymous, are Manichean dichotomies about what is true and what is false, what is real and what artificial. Of particular relevance is the Freeman emphasis on physicality.

FIGURE 3.   Anon holding a Union Jack defaced with Freeman discourse (photo by Jordan Mansfield, Getty Images)

Freemen make a hard distinction between the natural self—their own bodies and minds which are referred to by their first names and subject only to common law jurisdiction—and the "fictional" person, a legal construction that is used against them. In a dramatic and emotional denouement to the lecture, Harris calls for a mass "sacrifice of the person" through the voluntary destruction of the birth certificate. This, he holds, will liberate his audience from their bonded state. I explore this dichotomy between the person and self further in chapter 4.

Patterson's influential theory of social death finds a vernacular analogue in these ideas. His associations between slavery, death, and the dynamic of what he calls "natal alienation" are all present.[30] Patterson introduces the concept of natal alienation to articulate the core symbolic mechanism through which the slave's social death is publicly achieved. Instead of inheriting rights and privileges at birth (like the citizen of a modern state as modally conceived), the slave is instead sundered from all natal ties, whether of ancestry, locality, or any other form of group membership, denying them any claim of birth while recognizing only one social bond—that of their ownership. For Patterson, it is this symbolic work of uprooting and retying that legitimates the master's jurisdiction over them. When Anons describe themselves as slaves or serfs, or consider their birth certificates as the means by which they are owned and commodified, they convey a sense of the condition that Patterson identifies across history, namely one of nonpersonhood relative to a regime of power.[31] While it is indubitably perverse for contemporary citizens of the United Kingdom to compare themselves so readily to the subjects of chattel slavery—when their physical and social experiences differ in so many concrete ways from them—the emic persistence of this comparison reveals a degree of alienation that should be interrogated and understood.

The second set of notions explored by Pete are much more widely known and have been the subject of a substantial literature.[32] They begin with a totalitarian vision known as the New World Order. The New World Order (hereafter NWO) can be defined as a world government of unelected power-holders who override all claims of sovereignty and citizenship to control the earth for their own purposes. The first mainstream use of the term was by George Bush Senior in a speech to Congress at the end of the Cold War, intended to harken a new era of international security and stability. Yet "unbeknown to him," Michael Barkun explains, "The new world order was already a well-consolidated element in the thinking of both religious millenarians and those on the extreme political right," for whom it possessed deeply malevolent connotations.[33] As a consequence the speech was immediately interpreted by these sections of the American population as a brazen threat to their existence, even leading to the formation of armed militias ready to defend themselves.[34] Such fears were consolidated into a best-selling book of the same name published in 1991 and have been amplified by many other popular cultural artifacts since that point.[35] As the 1990s progressed, the concept of a NWO developed into a kind of "apocalyptic lingua franca," with the capacity to transcend a vast plurality of theories about the nature of contemporary power.[36] In the wake of digitalization they became more prevalent outside the United States, partly because, as Birchall observes, they were already recognized as profitable by emerging platform companies.[37]

Although the term is popularized at the end of the twentieth century, the NWO draws on older suppositions of premillennial Christianity about the coming of the Antichrist.[38] Following John's Epistles in the Book of Revelation, premillennialists assert that the end of the world will come when this diabolical figure (or figures) arrives to secure the control of world for Satan.[39] Having ostensibly been exported from the United States to Britain, Barkun argues that the traffic of ideas that pre-date this concept originally moved in the other direction. Of particular significance is nineteenth-century British evangelical John Nelson Derby, who saw the end of the world in the form of a global dictatorship led by the Antichrist, which would finally be swept away in a great conflict preceding Christ's return. As theories about the Antichrist developed throughout the twentieth century, they became hitched to the emergence of twentieth-century technologies: first television, and later computers and microelectronics. Indeed, some premillennialists saw the Antichrist itself in the form of a giant computer, tracking the lives of the innocent.[40] It is thus particularly apt that discussions about the coming of the NWO are taking place outside Britain's communications monitoring headquarters, as technological surveillance has been a long-running theme in this strain of diabolical thought.[41] This poses a special conundrum for Anonymous with its characteristic cyberutopianism and is sometimes poised in terms of great poles of dark and light, in statements such as, "These tools that have been put there to control us are actually liberating us."

AnonUK is self-consciously secular. This means that, although as Stewart and Harding observe there is no consistent dividing line between religious and secular millenarianism, the particular conception of the NWO that Anons share draws on a trajectory that has configured the Antichrist in secular terms. In Christian doctrine there is dispute over whether the Antichrist is a single individual or an organization, and this secular trajectory takes the latter view, seeing this evil dominion as operating through a network of secret societies. Foremost among these is the Bavarian Illuminati, a clandestine antiestablishment group that was briefly formed and disbanded in southern Germany before the French Revolution. Almost immediately, fears of the group were nurtured in counterrevolutionary texts, and a mythology around it grew in complexity and stature over the course of the nineteenth and twentieth centuries—which included holding it responsible for both the French and Russian Revolutions respectively.

Besides the Illuminati, Anons cite other secret sects that appear in the historical record, namely: the Freemasons, the Knights Templar, and the Skull and
Bones.[42] The antisemitic dimension of this trajectory was established, according
to Barkun, in interwar England, when a number of Jewish financial families were
considered to form significant nodes in the network.[43] This secular antisemitism
has undoubtedly been reinforced by the religious antisemitism of some funda-
mentalist Christians, for whom the Antichrist takes Jewish form.[44] Although the
tacit antisemitism that has been sedimented in these theories is unequivocal, it is
important to clarify that those in AnonUK—some of whom describe themselves
as antifascist—would reject this description wholeheartedly. Indeed one of the
cited explanations for why AnonUK is so avowedly secular is to distance itself
from histories of religious persecution.

As an apocalyptic lingua franca, the NWO are associated with many millennial theories, and Anons broadly adhere to one known as Agenda 21. The term
Agenda 21 also emerges in the early 1990s, as the title of a UN resolution to
combat climate change. For Anons, however, Agenda 21 is understood to be a
plot by the NWO to depopulate the earth to a maximum of five hundred million
inhabitants. Evidence for this project is provided by a mysterious piece of public
sculpture in the American state of Georgia, known as the Georgia Guidestones.
The Guidestones were originally commissioned in 1979 by an unknown man
working on behalf of a secret organization, of whom there remains reportedly
no historical trace, creating a compelling air of intrigue.[45] Dubbed the "American
Stonehenge," the assemblage consists of four sixteen-foot granite slabs poised in
a star shape around a central cuboidal pillar, all mounted by a large capstone.[46]
On the slabs are engraved ten "instructions" delivered in twelve languages, one
of which reads "Maintain humanity under 500,000,000 in perpetual balance
with nature." This inscription is interpreted as concrete evidence of Agenda 21:
an association first made publicly in 2005, and subsequently amplified by Alex
Jones, an influential theorist of the NWO.[47]

The specter of future death at the hands of a malevolent They is alive in the
minds of Anons. This is not mere dystopian pondering but can have material
consequences. In keeping with Pete's jacket, however, these tend to be defensive and individually oriented. Several Anons carry around survival kits or "grab
bags," which consist of emergency medical equipment and food supplies. One
mother of three provides each of her children with their own grab bag containing
sleeping and eating equipment, rather like a camping kit. Other male Anons discuss purchasing disused bunkers underground so they would have somewhere to
escape to—although something like this that would require larger resources generally remains more conjectural. The constant flow of crises evidenced by images
and videos online—ecological disasters, panicked crowds, empty supermarket

shelves—provides continual nourishment to these fears of end-times. It is hardly surprising that during the Coronavirus pandemic of 2020, Agenda 21 appeared to be in full swing.[48]

The Freeman theory of legal enslavement, and the theory of depopulation, contradict one another if we view them as political projects. If, however, we view them as vernacular expressions of a shared occult cosmology, they become two sides of the same circle. Both attest to the existence of unseen and mysterious powers who command the great forces of life and death. The holism of these theories should, I offer, not be downplayed. As detailed further below, Agenda 21 can extend to the air they breathe, the water they drink, the skies, and indeed in an occasional ambivalence around computing. Through these theories they share with other cult forms in accepting the implied existence of a supernatural.[49] It is nonetheless a dark supernatural; rather than one which will save souls, it is one from which souls must be saved. Because of the central importance of the supernatural, Stark and Bainbridge argue that cult movements are more likely to flourish in conditions of secularization. Where people already possess developed supernatural concepts, those that arrive with and through the cult are less able to successfully establish themselves. The broad secularism of key participants is consistent with this proposal. What may be added to this argument is that the Christocentrism of many of their biographies may lend itself much more readily to secular theories of the supernatural inflected by Christian theology. But what are the continuing consequences for this battle between good and evil at the protest in Cheltenham?

Humor and Nonviolence

The protest is scheduled to last the entire weekend, and while en route on Saturday I receive a text from a participant asserting that there will be a better turnout today and tomorrow. He is right. Conditions are more favorable than the previous day; it is warmer and the sun is out, and upwards of thirty people are there at its peak. The atmosphere is different, too, with a leisurely feel replacing the previous day's austerity. Friday's tinny smartphone has been substituted with a small sound system, and half a dozen tents have been erected in the grassy area beside the road, by those who plan to stay the night. Around lunchtime, the activists are collected into small porous groups, talking and laughing among themselves, when one man decides to bring out a megaphone and offer focus to proceedings. He begins with great solemnity. "At the moment there's a mockery of democracy, because we have a government agency called GCHQ that spies on everyone illegally." Yet almost as soon as he has started, the inclination for humor becomes irrepressible.

> So we might make a rockery for democracy. If you've got a rock, and you want to put it over there, bring it along. We might even get some crockery, and throw it, and do crockery for democracy. *(Raising his voice in a headmasterly tone)* If anybody's got any crockery, that's always good to throw. Not at anybody I hasten to add, but just generally in the direction of that way *(indicating toward the building).*

The joke catches, and the attendant Anons turn amiably toward him with expectant faces, tittering along in amusement. A younger man with a blazer full of badges and a shock of dyed blonde hair moves into the mood produced, taking up the megaphone next. In an impersonation of Alex Jones, he assumes a gravelly American voice and tilts his head slightly backward as he speaks, pushing out clauses like jets of water.

> You know the NSA doesn't stand for National Security Agency. It stands for Naughty—Saucy—Adults *(chorus of chuckles)*. All these adults? They're hiding. And you know what they're hiding from you? They're hiding the god-damn cake! And I've been looking for this cake! And they're constantly telling me, "The cake is a lie." Well I put on these pounds, and I put them on looking for this cake. So someday, those Naughty Saucy Adults, they will give me my cake. So, I say, we walk all the way, down the road and we go and find a café, and we start our search for the cake. Who's with me? *(Amused shouts of "yeah!")*. *(He changes his accent to Scots in a reference to the film Braveheart)* They may take our lives, but they will never—take—our cake! *(Cheers).*

This short speech is among the places where AnonUK comes closest to the individuation examined by Coleman. In the form Anonymous took online between 2006 and 2013, she observes the central importance of humor as a binding value and a praxis. Collected beneath the term "Lulz," a twist on the acronym for laugh out loud, humor here establishes dividing lines between insiders and outsiders, between those who get the joke, and those who are the object of it. What particularly characterized the Lulz in this cultural precursor was the presence of a victim. A praxis of online activities frequently enveloped in extreme speech, the Lulz explores the intimate and subtle relation between humor and violence—with violence taking an exclusively symbolic form.[50] Here the man with the shock of blonde hair is applying an innocuous version of what the Lulz can be. The term he uses, "the cake is a lie," is an idiom meaning that the promise of a reward is only a fictional motivator, originally popularized through a computer game.[51] The implication is that protesting against something as large as information-gathering is fruitless, an implication he jokingly both rejects and accepts, suggesting they go and look for the cake, a journey that would remove them from the protest itself. The reference to Braveheart represents his audience as valiant underdogs, centering the audiovisual imaginary, in this case historical fiction.

After this the megaphone goes quiet. Yet the joking energy it has released seeps through the quiet, hunting for new hosts, and after some hushed commotion in the tents, three women surface holding a blue plastic potty with a yellow liquid inside it.[52] They carry it toward the gates of the facility, with swift strides and suppressed giggles, and are soon apprehended by two unsmiling police officers. The women greet them, asking them where the nearest place to the building might be to empty the contents of the object. One of the policemen asks them, with apparent sincerity, what is in it. "Wee," the woman carrying it responds, deadpan. A second submerges her fingers in the liquid and proceeds to lick them in a gesture of pleasure. By this point another Anon has jogged onto the scene, immediately offering a satirical narrative to what is happening. He makes a short speech about how those inside the facility have been "taking the piss" (an English idiom meaning treating someone without respect), and therefore, by way of a mimetic reply, they would like to present the substance back to them, in the form of a trophy. He asks the five or so now assembled to give the facility a round of applause, and they duly clap in mock obsequiousness. As a fitting conclusion to the speech, emulating as it does an occasion dinner or an award ceremony, he proposes a toast. Except where, at the latter, everyone would raise their own glasses and drink, here the invented ritual begins to resemble something more like communion. The potty is passed around the small semicircle, each participant taking a sip. When it reaches the end of the line, another man concludes the narrative with mock jubilation, "A toast to GCHQ!" he cries, and with two hands tips the remainder of the vessel into his mouth.

What exactly is happening here? The revelations and concerns that ostensibly stimulated the protest are serious ones, and yet this small piece of theatre is entirely the opposite. Christena Nippert-Eng's concept of "boundary play" offers a helpful insight.[53] Boundary play is the visible manipulation of cultural categories, particularly through space, for the purposes of amusement. It can be recognized when things are not where they are meant to be (such as a potty at the gates of a high security facility), and this out of place quality somehow imparts laughter or pleasure. It is particularly likely to occur, she argues, in environments where categorical dichotomies are strong. Nippert-Eng develops this concept in contrast to another she calls "boundary work," the more somber and ongoing effort at drawing categorical boundaries that is essential to meaningful cultural life.[54] The distinction between boundary play and boundary work in online cults has important implications for violence. Consider, for example, an event that followed several years later in January 2021, when a number of activists, some of whom were associated with QAnon, breached the US Capitol. Though this episode demonstrated elements of "fun"—with a variety of forms of costumery—it was not play as such, but a form of work with concrete targets. The breach came in the wake of months of explicit calls for violence on IRC, and resulted in the injury of 140 police officers, three of whom died, as well as the deaths of three participants. While the protest outside GCHQ is taken seriously by the security services, involving the expenditure of substantial visible resources, and despite the protest being presented online as a confrontation between We and They, in the final analysis what is more important for those present is the We's collective self-formation. Through this form of play they explore and nourish their own humanity and togetherness.

This chapter began with three pronouns that arise in the AnonUK worldview. It remains only to elicit their associations with kinship. The We/They dichotomy is sometimes represented as a kin/non-kin distinction. Beneath these proper nouns, They are sometimes thought of as a small kin network, an endogamous set of "families" and "bloodlines" with malevolent intent toward all those outside them. On the other side, We are also represented in kin terms, as an "AnonFam" or as fictive siblings.[55] The shift is that We are moral, and therefore human, while immoral They are inherently inhuman. The effect is to rewrite the categories of humanity, with these new distinctions becoming much more important, at least on the surface, than other forms of segmentation such as nationhood or class. As with any other assertion of categorical difference, Anons continually face the ambivalence and confusion that arises from their porosity or contravention in practice. Before confronting the ambiguity of the You (who may be their actual kin or affines) I turn toward another pronoun which possesses its own generative schism—the I.

4

MASK

The mountain . . . told us to cover our faces so we would have a face. It told us to forget our names so we could be named.

The Zapatistas (1998, 22)

The presence of masks in situations relating to transition is so commonly the rule that exceptions to it are hard to find.

A. D. Napier (1986, 16)

In the southernmost reaches of South America, two Selk'nam boys undergoing a puberty rite enter a ceremonial hut. A large fire burns in the center, and initiated men are seated all around it, silent amid the shadows. Without warning the men suddenly start to chant loudly, and the initiands are told to stare into the fire, their skin glowing red with paint. From out of nowhere two masked spirits seem to rise from the flames and set on the boys in a full body fight. The boys are permitted to fight back, but they are strictly prohibited from removing their masks. After about half an hour both of the spirits kneel down, and the men inform the boys that the fight is over. They point to the mask on one of the spirits. "Touch it!" they order.[1] Still fearful and quivering from their exertions, one of the initiands slowly reaches out and lays his hand on the mask, tentatively pulling it off the spirit's head. Revelation spreads. The face belongs to a man he knows well—and the man is smiling at him.

Among some of the earliest cultural artifacts, anthropologists have been considering the social significance of masks since the discipline's beginnings. For much of this history of study, the focus has been on what a mask *means* within a given society. Encompassed within broader concerns with art, ritual, and myth, masks have been assessed as part of wider symbolic systems that it is the anthropologist's task to decode.[2] Yet Anonymous's signature mask has always obstructed this kind of decoding. It has never been straightforward, to neither journalists nor scholars, what exactly this mask means for those who associate themselves with it. This obstruction has so far been circumvented by turning the concept of meaning inside out. Drawing on poststructuralist philosophy, Lewis Call describes the Anonymous mask as a "free-floating signifier" whose power resides in its ability to resist any enduring attachment to a signified.[3] As a symbol that negates the very principle of symbolism, it becomes a malleable tool of political subversion. Coleman adopts a similar position, describing it as standing for "everything and nothing at once."[4]

In this chapter I approach this now famous mask differently, by drawing on a different intellectual genealogy. Since the 1970s, greater attention has been paid to how masks are implicated in active social processes. Ethnographers began to document in more depth how masks were made and used as part of fertility, medicinal, and political rituals, carnivals and masquerades—and initiations.[5] Beside this shifting ethnographic focus a performative theory of masking developed, interrogating how the wearing of masks can open up a new relation of the wearer to themselves and to their audience.[6] It is this active, performative approach that I adopt here, one that asks more directly what the mask is doing for those who decide to wear it.[7] As a performed object, it is normally worn above the head, thereby displaying the face of the activist. Comparable to the revelatory moment above that Michael Taussig describes in a Selk'nam male puberty rite, in these instances the mask indexes a moment of unmasking, peeling back conventional representations like a skin being shed, to display their new identities as awakened.

In her ethnography of Alcoholics Anonymous and related groups in Portugal, Catarina Frois explores the way anonymity functions as an "operative concept" therein.[8] The value of anonymity, she argues, does not actually rest in being anonymous in any meaningful sense of the term, but in enforcing productive boundaries between different parts of members' lives. By distancing themselves from legal and family names inside the meetings, it allows for the establishment of distance from stigma, and thereby a transformation from stigmatized to normalized forms of personhood. It also (a phenomenon that equally occurs when initiands lose their former names) allows for the collapse of structural differences between members in which names are perpetually implicated, producing a powerful sense of sameness.[9] Anons' performance of unmasking is sometimes echoed in comparable forms of legal name avoidance. Through it, they assert distance from a bureaucratic apparatus that is experienced as dominating, avoiding the summons in which legal names are implicated, and the obligations that come with it. Below, I explore these variegated connections between masks, names, and the austerity state, through a march held for the homeless in Manchester. In particular, I focus on an instructive anecdote relayed by a young participant called Simon.

As the tipping point between the two parts of the narrative—death and waking up—this chapter stitches together a number of theoretical arguments that have already been set out. Building on the experiences of separation chronicled in chapter 1, and the modality and form of their theorizations described in chapters 2 and 3, it begins by exploring the Guy Fawkes mask as a stand-in for the prevailing sense of invisibility that Anons share. Putting on the mask, like deciding to join AA, is a liberating moment of acknowledging a stigmatized identity, which is to say, being invisible. Yet like the latter, it equally contains the potential to produce the radically opposite effect, in this case becoming highly visible in the British public sphere. Rather than masking, then, it is consequently *unmasking* that is the central dynamic and project of the cult. Culturally speaking, it is one underpinned by inherited post-Lutheran and state-oriented ideas about what constitutes the self, and how it can legitimately be changed.

The Solidarity of Invisibility

Anons identify one another by the wearing of a mask. The normal form of the mask is made inexpensively from plastic and can easily be bought online for a few British pounds. There are no rules about who can wear it. It remains a core principle in AnonUK, continuous with prior individuations, that in theory anybody retains the right to wear the mask. With black bushy eyebrows, a moustache, and a morally ambiguous grin, the face it represents is nominally that of Guy Fawkes and is a reference to the film *V for Vendetta* in which it is worn by the Guy Fawkes character. Its specific contemporary form can be attributed to David Lloyd, the illustrator of Alan Moore's eponymous comic upon which the film is based. This peculiar combination of eyebrows, moustache, and smile, however, pre-dates even Lloyd's design. Similar iterations are in evidence in eighteenth-century Europe, while fig. 4 shows a mask from Western Iran or Anatolia with remarkably similar features. While a visual history of the Guy Fawkes mask is beyond my aims here, a record such as this suggests its earliest origins may be Asiatic.

The mask can produce contrasting reactions in its audience. For people who do not self-identify as Anon, the sight of it can be disturbing. It is variously described as "sinister," or with "undertones of menace." Some are reputedly "scared" of the mask—a response which follows in the long tradition of initiation masking, in which such masks are deliberately worn to terrify the uninitiated.[10] For Anons, by contrast, the reaction is entirely different. They describe intense and spontaneous feelings of trust that the mask is able to inspire in its wearer. This may lead to their immediately exchanging e-mail addresses or Facebook

FIGURE 4. War mask, c. 1450–1500, Western Iran or Anatolia (photo by Marc Pelletreau, The Museum of Islamic Art, Doha [MW.6.1997])

monikers, or even engaging in a two-armed embrace. When they endeavor to describe what lies at the root of this instant trust, Anons frequently reach for some expression of sameness. The mask is an "equalizer." It means they are there "for the same thing" and possess "common cause." As a signifier it is experienced as something that effaces the difference and distance between those who have decided to wear it, producing a profound sense of assembly. As one says, "Whatever protest I go to, the one place I'll always be at home is behind that mask, because for me the mask means one thing and that's solidarity."

This way of talking about the mask and the feelings it evokes recalls what Victor Turner called "communitas," Latin for community.[11] Turner employs the Latin to distinguish it from the English word community, which usually implies a geographical area, while articulating a communal way of relating that infuses the condition of liminality. He defines communitas as "a matter of giving recognition to an essential and generic human bond, without which there could be *no* society."[12] Like the sentiments aroused by the Guy Fawkes mask, this bond is characterized by sameness, equality, and a lack of status difference.[13] Communitas plays an important role in Turner's work as a term which represents the opposite of structure. If structure is defined as an arrangement of roles and ranks that differentiate people from one another (such as age cohort), communitas dissolves such categories, and it is through it that new connections can be made. Indeed, in his later work, Turner acknowledges the debt his concept of communitas owed to his study of African puberty rituals, having observed how Ndembu initiands formed friendships during transition that were not determined by their neighborhoods or family ties.[14] In traditional rites like these, the dissolution of status allows the ritual subject to change status, transforming them from one role to another. In his later focus on large-scale industrial societies, Turner saw the sensation of communitas as a major engine of social and political change.

Equating the Guy Fawkes mask with the leveling dynamics of communitas, is quite apart from the role masks have played in asymmetrical settings. In the latter, masks have been indicators of personhood, the right to own or wear them a sign of elevated role or status. As Lévi-Strauss makes clear, many types of ritual mask are considered high value objects, passed down through noble lineages in hierarchical societies, or available for purchase in socially mobile ones, endowing their owners with value in each instance.[15] More saliently, undergoing a traditional initiation often endows the newly initiated with the right to wear a particular mask. In some cases, this right will be granted only to those who have made some show of merit during the initiatory ordeal, while in others the ordeals themselves are staggered, with each new rank of initiation bringing with it the right to wear a higher ranking mask.[16] Indeed, the crowning moment of these rituals often comes when initiands emerge from their seclusion, wearing the masks that can be worn only by initiated persons, walking or dancing them in full view of the community.[17] The mask confirms their sanctification by the ritual, and with it their access to new roles, such as taking part in religious rituals or having the right to marry.

In these examples wearing a mask entails becoming "someone." In AnonUK, wearing its signature mask is, in some form or another, to embrace being no one. By rejecting the face around which personhood in Britain can pivot, this form of solidarity assembles around a negative rather than a positive identity. It is not that kind of solidarity that obtains between people who share a specific relationship to the means of production (such as a peasantry or industrial working class); nor is it the solidarity that obtains between people who share marginalized ethnic, gender, or sexual identities. Instead of the bond that comes from possessing a role in society, it entails uniting around the absence thereof. A profound subjective knowledge of invisibility binds Anons together, despite each having been individually manifested. It is worth observing that a shared negative identity can be felt to be equally as powerful as a positive one.[18] Turner and others have witnessed the lasting alliances that can grow between initiands in the liminal state, forged in the furnace of hardship.[19] In AnonUK this takes a more enduring form, one we might think of as a solidarity of invisibility.

The argument would end here were we to rest on the mask's meaning. However, if we consider it as a *performed* object, a new layer of ambiguity settles. A. D. Napier documents the long-standing association between masks and paradox—the proposition that two contradictory things can both be true—and this paradox is particularly acute in the way that Guy Fawkes masks are most often worn.[20] Contrary to popular conceptions that surround this individuation of Anonymous, most Anons do not wear these masks over their faces in order to conceal them, but instead prop them up on the crown of their heads, revealing their faces in the process (see **fig. 5**). This particular performance is one among a diversity of masking practices (see C in **fig. 6**) and has been referred to by anthropologists

FIGURE 5. Three activists wearing raised Anonymous masks at Million Mask March (photo by Lucas Somavilla Croxatto, anonymized by Benjamin Elwyn)

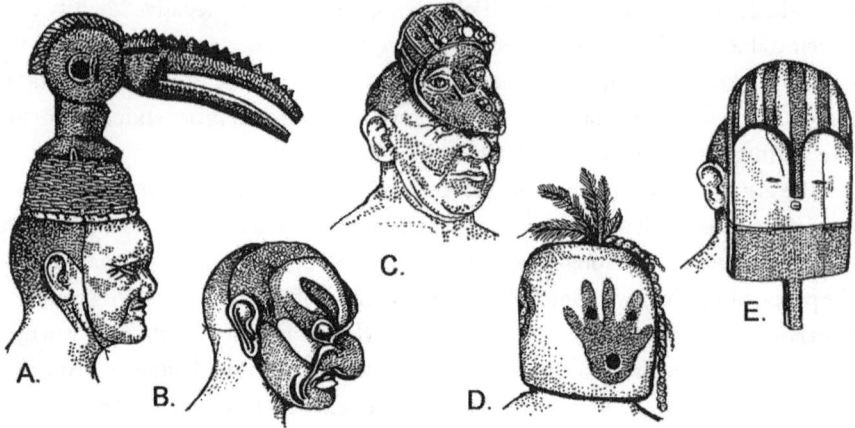

FIGURE 6.   Drawing of five different ways of wearing a mask (by Gary Edson, McFarland)

as a "forehead mask." A forehead mask has a separate effect from a mask that covers the face or head, as it visualizes a rupture in the wearer between two ways of being. In the case of AnonUK, it signals a symbolic death that is being cast off, while demonstrating a new fleshy life that persists as a generative force just beneath it.

Performing Unmasking

It is a blanched, watery midday in Manchester. The crowd is slowly swelling in Piccadilly Gardens, a square in the heart of the city with strips of grass and a white Ferris wheel. The demographic is mostly thirty to forty-plus, who besides casual passersby display a range of affiliations—advocates of a homeless charity, supporters of a would-be independent MP, a samba band—and dotted around, at least a dozen Anonymous masks. The hundred or so have gathered to attend a march for the homeless taking place that day in a number of cities across England, which has been heavily promoted online in several Anonymous channels.[21] Here in Manchester, the administrator of its Facebook page is a self-identified Anon— an angular White man wearing dark jeans, trainers, and a burgundy sweatshirt embossed with Anonymous iconography in gold. On the crown of his head sits the slightly jaundiced version of the Guy Fawkes mask, fixed to the nape of his neck with a band of black elastic.[22] He kicks off proceedings with a short speech, citing the effects of austerity on the city.

> This month Manchester Labour Council are cutting back funding from mental support services and homeless services . . .[23] With these austerity measures coming into play with the universal credit, the bedroom tax . . . People are dying. I'm here every Saturday with my amazing team, feeding the homeless, and we're feeding *hundreds* of people.

Other Anons scattered throughout the crowd watch on. They are both male and female, and are mostly White and appear to be over the age of thirty. Besides the Guy Fawkes masks which are propped at similar forty-five-degree angles above their faces, some of them also hold up homemade cardboard signs. The messages they contain articulate an identification with homeless people. It could be you, they suggest.

How many pay cheques could you afford to miss?

You are only two pay cheques from being homeless

Two diffident men stand at the margins, clasping their masks in their hands. They wear similar branded Anonymous clothing to the speaker. When asked what the symbolism of Anonymous means to them, one expresses this negative solidarity.

> Basically making people stand together and show their support. Helping the homeless, helping the needy, helping the sick. People that are being pushed around, pushed aside, being told no all the time when it should be yes.

He bought his own mask after the introduction of the Bedroom Tax—a law passed in 2012 that entailed a reduction in benefit for council housing tenants with spare rooms.[24] It affected him and it affected his mother who was disabled and mentally ill.

When the march starts the pace is quick. The vanguard led by the man in burgundy are practically sprinting. This is not the steady patient tread of a trade union or antiwar demo. In fact, it does not seem to be about marching at all, but about reaching the desired destination as quickly as they can. They achieve this in just a few minutes, as people start to pour into the street outside Manchester Town Hall, an opulent neo-Gothic structure that speaks to a former era of grandeur investing the organs of the British state.[25] Once here the atmosphere turns. Rather than a plaintive plea against the cutting of state budgets, there is a more vigorous desire for confrontation with the Council. The man at the front leads the charge. "Come on!" he impels those around him, his angular face taut with urgency. Most hang back, but a few dozen join him in attempting to enter the building by way of a narrow stone staircase. As they ascend they are uncomfortably compressed, bodies pushed tightly against each other, limbs bent at unseemly angles as their faces contort into grimaces and groans. Eventually most are prevented from entering the building by a line of security guards, and an ornate wrought-iron gate that is slowly lowered, as a participant jokes, "to keep the peasants out."[26]

This instigates phase two, which is to camp directly outside it. A number of Anons make their way around to the side of the building, where there is a spacious paved square, and the machinery of a homeless solidarity "sleep out" whirs into action.[27] Tents are flimsily pitched on the cobblestone ground, and a food stand rises where women with raised Guy Fawkes masks serve the homeless free hot dogs and drinks. The skirting of the building and the surface of the square are rapidly inscribed with chalk messages, and on the opposite side, beneath an imposing statue of Prince Albert, someone holds up another piece of cardboard that reads, "We are the voices of the voiceless."[28] By this point the initial confrontation has given way to a slower-paced presencing. They position themselves in a key location facing the main entrance of the Town Hall, where it would be impossible for its occupants to ignore them.[29] As the mood settles, a small group begins to chant in unison, "What do we want? End the austerity! When do we want it? Now!"

Standing next to the chanting group is Simon. He is a twenty-three-year-old man, with pale skin and eyes, and an approachably benign demeanor. Simon's "lightbulb" moment came in 2010. He was eighteen that year and on the verge of going to university, when university tuition fees in Britain were raised to £9,000 a year, altering his decision. When he first woke up, he remembers, he was "angry": first at having been "fooled for so long," and second, "because people I really cared about still couldn't see it." Simon resolved that the best way to change the system that had steered the course of his life in this way was to abandon the system, patiently disassembling it with its own apparatus. He took it upon himself to apprentice into aspects of British law, and he talks fluidly about esoteric legal instruments such as promissory notes and tripartite contracts. Now he runs a Facebook group with more than twenty thousand members that helps to prevent people from being evicted from their homes. He continually stresses the importance of nonviolence, however, and maintains the value of "educating" others to secure lasting change. This extends even to the bailiffs. "They get schooled," he says with assurance, and then, "They'll walk away a different person."

One of the concrete results of Simon's awakening has been to boycott certain taxes, such as Council Tax and the license fee that funds the BBC.[30] Simon describes, with some relish, an encounter he once had with a representative from the TV Licensing authority, who arrived at his door to check he had the right to own a broadcast television.

> I opened the door and they said, "Can I speak to Mr. Roberts?" Who the hell is that? Who the hell is Mr. Roberts? That is a legal fiction that. That is a legal title. That is not a physical human being! (*His tone rises with indignation.*)
> *To the question of what he would rather be called, he replies,*
> Just Simon. (*He pauses briefly.*) Roberts. So they go, "Are you Mr. Roberts?" And I go, "Who the hell is that?" And they go "Well I've heard that he lives at this address." And I went "He?" I went, "A legal fiction? You know I didn't know legal fictions had gender." (*He chuckles at the memory.*) And they're confused. They don't know what you're talking about.

One can imagine the puzzlement forming on the face of his interlocutor, on meeting someone so outraged at being addressed as Mr. Although Simon asserts he does not identify with the Freeman movement, here he is rehearsing some of its key precepts. In these, his legal personhood (or "fiction") that is created in the moment the birth certificate is drawn up ("Mr. Roberts"), becomes something alienated from him as a physical human being ("Just Simon"). What is instructive about his retelling is the passion underlying it. It is not simply that Mr. Roberts has become alienated from Simon, but that these two figures now exist in a relationship of mutual antagonism. As the representative of his politico-juridical existence, Mr. Roberts is not only separate from Simon as a physical body, but is also felt to be coercing him through their ongoing and inextricable association—like a jailer who holds the only set of keys to his cell. Although not normally expressed with such fervor, Simon's story is emblematic of a wider critique of personhood that prevails in Anonymous. They consistently frame personhood—sometimes even using this word and its cognates—as something derogatory and inauthentic. For instance, someone alienated from their own humanity is described as "lost in the person."

Through these distinctions, Anons rehearse in an extreme form, a much broader set of assumptions that exist across Euro-American societies. Marcel Mauss offers a meta-analysis of this way of thinking in his influential essay on the emergence of the idea of the self.[31] Mauss's principal argument is that contemporary notions of selfhood (what he calls "just Simon"), have their origins in ancient notions of personhood, namely of inhabiting specific social roles, that were codified in Roman law and reproduced through legal institutions. Following Mauss, the paradox above is that "just Simon" is in fact potentiated by the very existence of "Mr. Roberts." Critical to leveraging this long transition from something social and specific, to something individual and universal, are in Mauss's view the effects of a post-Lutheran Christianity, which sought to place the individual soul in a direct relationship to God. Indeed, this essay brings us full circle, as he observes that the word person derives from the Latin word *persona,* meaning mask. Though he does not elicit this aspect, the genealogy he draws has proportionate implications for cultural assumptions surrounding masks that ramify into the present. Through this transition from valorizing the inhabiting of a social persona, to valorizing being an individual, the mask develops alongside into something synonymous with fakery and illusion. This is particularly notice-able (and influential) in early twentieth-century psychoanalysis, where the mask comes to stand for a way of being that hides the true self.[32]

The story of Simon's encounter with the licensing authority finds its symbolic analogue in the performance of the forehead-mask. For if the mask is equated with personhood is equated with illusion, and the face is equated with the self is equated with truth, then raising the Guy Fawkes mask can be understood as an attempt to dissociate themselves from something illusory, and through this, to reconstitute a sense of themselves as authentic.[33] It is through their performance of the Guy Fawkes mask that the paradoxes that surround the attitude UK Anons actually have to anonymity become starkest. Anons come to protests, they say, to "show" their faces; they try to be Anonymous, so people "know who they are."[34] I now consider this phenomenon in comparison.

Thanatoid Masks

Taussig develops the idea of unmasking through the concept of "defacement."[35] A practice that frequently centers on the revelation of the face (literally de-face-ment), it involves the desecration of a public symbol, to redraw distinctions between the true and the false, and thereby reconstitute the real. In European modernity there are countless examples of this dynamic, but it is not confined to contexts that draw on European ideology.[36] Taussig spends a significant amount of time exploring the concept from Isla Grande, a large island at the southern tip of South America, and in particular during the Selk'nam male puberty rite. Here I would take Taussig's tacit emphasis on initiation much farther, because in many accounts of the ritual, comparable episodes of defacement occur. The most com-mon genre is the one he retells, in which the faces of initiated men are deliberately revealed to initiands, to show them that the spirit world comes alive through the active work of human beings.[37] But in this literature defacement can come in two further forms. One is the covering and revelation of the face of the initiand, who may be masked or blindfolded and forced to undergo painful procedures in this condition.³⁸ The other is the defacement of a third party, normally an animal, who is ritually sacrificed and part of its head or face removed.³⁹ In each case, the moment of defacement constitutes a crucial transition.

Comparing this practice across a variety of contexts, there is a genre of mask-wearing that has so far escaped extensive attention—those that are specifically used in order to reveal the face. Unlike other masks that conceal the face or head in its entirety (see **fig. 6**), this way of wearing a mask pries open a material gap between the mask and the face, and consequently between the role and the human being performing it, and in this way visualizes the ultimate separation of their social and biological existence. Initiation is a process of becoming symbolically dead while remaining biologically alive, and these defacements happen during the period of symbolic death, as part of a wider effort to inaugurate the ritual subject's new social life. But all rites of passage turn in some sense around these paradoxes, and the practice can be found across the category. The moment at which the bride's veil is lifted during the modal Euro-American marriage rite, revealing the face to the beloved, is another example. Funerals are a particularly rich site for the masks that reveal. Masks and other kinds of face and head coverings have been used across place and time to signify the persistence of personhood beyond the grave, demonstrating in their physical detachment from the decaying body that the latter does not rely on individual biology to endure.⁴⁰ Indeed, the oldest masks ever discovered are thought to be death masks.⁴¹

As this genre hovers around experiences of social or biological death, I call these masks "thanatic" or "thanatoid"—adapted from Thanatos, the ancient Greek god of nonviolent death. The distinction between the two terms is initially inspired by Turner's distinction between the "liminal" and the "liminoid."⁴² Turner became known for his early discussions of liminality, and in his later work he sought to transport this concept beyond the space of its ethnographic genesis. He coined the term "liminoid" to theorize those segments of postindustrial society that stretched or even upended normal social categories, and therefore held the potential to be culturally transformative, the same way that the liminal period transformed Ndembu initiands. Turner's distinction between the liminal and liminoid tracks a distinction he offers between closed and open systems, i.e., between societies which have a developed sense of boundaries and those that do not. For Turner, liminal phenomena take place in closed systems, in those groups that maintain strong insider/outsider divisions. Along with the Ndembu, in this category he includes sects, clubs, and religious orders. Liminoid phenomena, by contrast, take place in open systems, among any group of people who maintain a porosity of membership. The dichotomy also broadly tracks a distinction he makes between pre- and postindustrial societies. While continuing Turner's project of exploring liminal transformation in radically different settings, it is this externalist approach that I depart from here. As a phenomenon, AnonUK clearly troubles these dichotomies. Entry is closed to the extent that participants must have "woken up," but open to the extent that "anyone can be Anonymous." It flourishes in a context that is postindustrial, but also increasingly de-industrial. Instead, I turn to Handelman to consider the internal dynamics of unmasking in each instance.

Handelman opens up the category of ritual by inverting its anthropological significance. In a critique of Turner, as well as Clifford Geertz, Edmund Leach, and others, his writings stand against a precept he describes as "near canonical in anthropology," that "ritual is a storehouse of culture and society, epiphenomenally shaped to . . . reflect on the latter."[43] This orthodoxy, he argues, carries three assumptions with it. The first is that ritual is always a representation of the wider social context in which it is occurring. The second is that ritual is functional for the order of this context, closely related to the first in that it plays a purpose that is externally defined. Finally, ritual is an arena in which social competition and conflict are played out. The fresh perspective that Handelman offers is instead to focus more acutely on the internal logic of ritual processes, and from *that* point assess what their significance relative to wider nets of social relations may be. By way of Handelman, then, we can approach the different performances of unmasking discussed here on their own terms, without first asserting what they may mean, or who they may be for. In the process it becomes clearer that the main difference between the way masks are used in traditional initiation rites, and the way they are used among those digitally initiated into AnonUK, hinge on diverging ideas about sequence. I gather some of these threads beneath the terms thanatic and thanatoid.

A thanatic mask is the mask used to unmask in a diachronic series of events. It is the visual focus of that illuminating moment that Taussig develops, when a mask is removed to reveal a face, producing a new sense of reality in the onlooker or the wearer, and a new way of navigating the world therein. What is important to note about a thanatic mask, is that it is deployed to produce a short-lived apotheosis as part of a longer ritual choreography, signaling the point at which social and biological life become decoupled, to make way for their re-pairing along a different axis. In traditional initiations, it is the moment at which a new form of personhood can begin to grow in the space left by the old and is normally succeeded by a process of re-covering (or "refacement" as Taussig calls it).[44] In initiation rites this means either re-covering the face with a mask, or in a broader sense, recovering the head and body with headdresses, cloaks, feathers, animal skins, or other forms of ritual ornamentation and attire.[45] Because the right to wear a mask, in its broadest definition, equals access to certain forms of personhood in the ancient sense that Mauss described, these coverings mark the completion of the rite.

A thanatoid mask is the invocation of unmasking as a metaphor, whether used visually or verbally. This invocation resembles the use of a thanatic mask, as it suggests the transformations that thanatic masks produce in rites of passage; however, it does not occupy a transitional location within a longer choreography. As a result its appearance does not necessarily need to be a fleeting event (although it may still be used as such), but rather can operate as a prevailing articulation of, or provocation for, change within a given social sphere. The way the Guy Fawkes mask is used by UK Anons in Manchester and elsewhere, when they wear it as a forehead-mask, can be considered in this genre. Sporting these masks in such a way as to reveal their own faces becomes not only an identification with Anonymous and all it may signify to other members, but is at the same time a performed reference to a moment in the past. While the thanatic mask proceeds diachronically through time—employed in rites of passage to mark and dignify the human life course—a thanatoid mask is a form of synchronic history. To put this abstractly, if a thanatic mask is moment C in a ritual sequence that goes A, B, C, D, E, then a thanatoid mask is the continual replaying of moment C (that is, waking up) in an iterative series that goes A, B, C, C, C . . . *ad infinitum.*

This counter-chronological approach to the use of the Guy Fawkes mask rehearses itself in larger ways in their attitude toward their rituals in general. For if those rituals examined by Turner, Gluckman, and others hinge on a particular temporal ordering of events, on the agreement that one thing should happen after another in a more or less orchestrated manner, then UK Anons would generally reject this. In Manchester, the patient process of ambulation that has historically characterized a political march is not there; all that really matters is their arrival at the doors of the council and their subsequent occupation of the square. Their principled refusal to direct the unfolding of time is apparent, to some extent, in all the rituals described here. Besides the start time (and even this is a movable feast), the public events that AnonUK administer tend to unfold in free-form ways rather than follow a planned itinerary. If they are organized chronologically, then this somewhat taboo form of work tends to be carefully hidden from view, and if it is made public, may become the subject of mockery.[46] Nowhere is this more apparent than at the "all-out chaos" of the Million Mask Marches. Indeed, one of the insider jokes in AnonUK is that the Million Mask March is not actually a march.

This brings us back to transformation itself, and where it is actually occurring. The difference between these unmasking processes in traditional initiations, and the idea of unmasking in AnonUK, arises from the different ideas about where personhood is made. In a much later article, Bryan Turner returns to Mauss's essay, to consider his question in expressly political terms.[47] He submits that the notion of a universal self that Mauss outlines arises not only through a set of theological and legal precepts, but also through the growing universality of the category of citizenship. His argument relates specifically to Britain. As the rights of citizenship were expanded to encompass ever-widening sections of the British population (working-class men, women, and later children and animals), and as these rights were increased to include a variety of employment rights, property rights, and rights to care, so too did an increasingly universalistic idea of the self. The rituals that matter in the formation of personhood are consequently bureaucratic rituals—getting a passport, receiving a birth certificate, and to a far lesser extent, paying the BBC license fee.[48] This is why rejecting the legal summons that arrives at the door is significant, as it is through these interactions that meaningful symbolic work is occurring.

The logic of a traditional initiation—broadly understood—is that personhood is conferred through its own procedures. There is debate among anthropologists about whether the men and women who take the form of metaphysical beings through the use of masks and costumes are really "faking it" or not.[49] Not only is ethnographic access to this kind of interiority challenging and somewhat problematic, but in an analytic sense it does not really matter. What does matter is the shared agreement that transformative planes of reality can be accessed through these artifacts, rendering them critical mediators of personhood changes. Indeed, this conflict between differing ideas about how personhood is made is not purely abstract. In the colonial period, Christian missionaries and bureaucratic European states converged in their castigation of ritual masking as idolatry and deceit. Masks were stolen, confiscated, and burned.[50] Even as late as the 1960s, initiation masks were still being destroyed in large numbers as part of modernizing "demystification" programs.[51]

Simon's passionate disavowal of Mr. Roberts thus constitutes—in a small way—another episode in this history of contestation. Though narrowed here for expository purposes, his story is emblematic of other tales of legal name avoidance in AnonUK. Sometimes this takes the form of multiplication: they may have several different names they use at their discretion for different bureaucratic and commercial transactions, or they may set up hundreds of different Facebook accounts (which enforces a "real names policy") that are subsequently closed down. At others it involves the committed assumption of a character—such as the Jesus figure who regularly treads at the Million Mask March. When Anons discuss their relationship to their legal names, it often takes the form of something which is what Susan Benson calls "injurious"—possessing the ability to harm them.[52] One puts this explicitly when he says, using the familiar adversarial pronoun, "They want a signature, they want a name on paper, so they can attack you." For this reason, Anons are respectful of others' rights to name avoidance, habitually employing the monikers offered on Facebook. Not once am I ever asked my name, and instead I tend to be interpellated, somewhat ironically, as V.

This chapter began by asking what the Guy Fawkes mask *does* for those who decide to wear it. Like the head with the mask above it, the answer gazes in two directions. The first is a statement on the self and the changes it has gone through. Here we witness some of the symbolic effects of austerity. For if, as Bryan Turner reminds us, symbolic value and politico-juridical value travel together, then the effective contraction of the rights of citizenship in Britain would entail a proportional diminishment of the symbolic power of the citizen as *persona*. One of the agreements of citizenship is that through your legal name you acquire rights and protection. This remains partly true for UK Anons—none of whom, in this ethnography are stateless; however, because of the structural carelessness they have experienced, they no longer see it this way. Instead, as with Frois's interlocutors, it represents a stigmatized identity from which they seek to distance themselves. The second part of the answer looks outward. Having gone through their digital initiations, Anons become impelled to initiate others. Every performance, at some point, needs an audience, and here this role is occupied, wittingly or not, by the You. Showing one's face facilitates these personal connections. It is the outward-facing effects of these rites that we turn to in the second part.

Part II
WAKING UP

5

KNOWLEDGE

Other people are like "there there." Anons, they *know*.

Anon, Nottingham

Two Bemba women are to be initiated in their elaborate puberty ritual—the Chisungu.[1] Their transition takes on particular importance as it is they who will be the start of new lineages. The ceremony lasts for over a month. There are long quiet periods of pottery making and drawing, interspersed with long loud periods of singing, clapping, and dancing to the drum. During the ceremony, the older women make and draw *mbusa*: sacred emblems which take the form of small and large items of pottery, and designs on the walls of the initiation hut. They have secret names and meanings that are taught to the girls in the weeks that follow, representing a variety of skills, attitudes, obligations, examples of good and bad behavior. Through the *mbusa*, "we make them clever," the older women say.[2] Yet the initiands themselves are hardly in a position to learn during the rite, covered as they sometimes are in blankets, or told not to look at what is going on. In practice they had already learned much, like mushroom picking, beforehand, and other things, like tending the grain, they do not master until years later. Presented with these simple images and figurines, they are taught how to think about what they know. It is this that "grow(s)" the girl.[3]

Waking up is represented by Anons as a singular rupture, a more or less dramatic episode of dreamlike insight. After this event, however, they normally relay far longer periods of digitally mediated learning, in which this revelatory moment is developed into a more comprehensive body of social knowledge. This chapter in some respects picks up from where chapter 2 left off. The digital initiand is now "awake," and in a more self-conscious process of knowledge acquisition.

Although, like many other contemporary online cults, Anons represent what they know, and how they came to acquire it, using the language of scientific rationalism, the quality of this knowledge and the method that led to it are different. Because the knowledge of invisibility provides the basis for theorizing, logically it cannot subsequently be refuted.[4] In this they align closely with other initiations, built around the deliberate entanglement of subjective and social knowledge. It is by effectively combining the two that they are able to join the knowledge bearing community of the We.

The centrality of knowledge to radical online groups is widely recognized.[5] While the popularization of alternative knowledge preceded the platforms where it is now cultivated and shared, there is little dispute that digitization has greatly intensified this shift, perhaps even beyond a tipping point into dominant epistemes of their own.[6] Indeed, the importance of knowledge in the ascent of the so-called alt-right, online assemblies concentrated in Europe and North America that promote forms of gendered and/or racial supremacism, is such that some scholars describe it as "knowledge activism."[7] This is particularly the case in QAnon. Q was so called because he or she purportedly held "Q level" clearance—state secrets at the highest level that were then disseminated as clues.[8] In these "Q drops," followers were encouraged to embrace the quest for knowledge, and indeed the word itself is used within them several times.[9] Like AnonUK, the alt-right draws heavily on *The Matrix* as a reference point. The scene in which Neo is invited to take either the blue pill of ignorance, or the red pill of knowledge, is repeatedly rehearsed in the expression "red pilling," an incremental online learning process.[10] AnonUK appeared before these darker potentialities had appeared, and thus they may be instructive in presenting this problem of knowledge at a more innocent historical moment.

Both of the major comparative studies of initiation center knowledge in their definitions of it. Eliade defines initiation as, "A body of rites and oral teachings whose purpose is to produce a decisive alteration in the religious and social status of the person to be initiated"; while La Fontaine asserts that, "All initiations purport to transmit knowledge and powers that are exclusive to the initiated."[11] In the course of these investigations, both scholars lay a slightly different emphasis on where they consider the significance of the rite to rest. Eliade suggests that knowledge sanctifies the person undergoing the rite, in which the main change is this greater approximation of the sacred. La Fontaine, writing some decades later, is critical of this kind of abstraction. For her the most important function of knowledge is to preserve existing forms of social authority, arguing that what is learned therein provides the tacit legitimation of that authority. I explore some of the implications beneath. For now, as a point of historical divergence, I would suggest that in digital initiation, knowledge plays a more outsized role than in its antecedents. Because the knowledge acquired online makes the defining slice in the nexus, it becomes even more pivotal to the sequence itself, showing in its wake that initiation has occurred.

This chapter unfolds the problem of knowledge ethnographically. It documents the ways in which Anons talk about the "truth" that for them lies inside Internet repositories, as well as their prevailing distrust of other mediums of knowledge transference. As elsewhere, I emphasize the significance of Christian theological precedents in providing the substance and logics of these discourses, specifically the doctrine of revelation, and the concept of "biblical inerrancy." As a narrative, the chapter opens with a march staged in solidarity with Julian Assange, before pivoting toward the perspective of one of its participants, Tank, and his particular view on the hidden nature of contemporary power.

March to the Embassy

An event is advertised on Facebook. It promotes a plan to walk from London's Trafalgar Square to the Ecuadorian Embassy in Knightsbridge, as a show of support for Julian Assange and Edward Snowden. Assange became widely known in 2010 when the whistleblowing organization he spearheaded, WikiLeaks, published a video documenting an American military helicopter firing on civilians in a Baghdad suburb, followed by a large tranche of American diplomatic cables. For the past year-and-a-half Assange has been living inside the Ecuadorian Embassy, where he was granted asylum to avoid an extradition order to Sweden and possibly the United States.[12] Edward Snowden is an ex-NSA contractor who, just over a year beforehand, himself released a large tranche of evidence that the American intelligence agency had been engaged in the mass surveillance of ordinary Internet users. What closely connects Assange and Snowden at this historical juncture, is a shared commitment to the moral project of whistleblowing. Echoing one of the core tenets of Free Culture—that information should be "free"—this project hinges on the conviction that if enough information becomes available, injustice and wrongdoing will be washed out of institutional life and morality refreshed. In expressing explicit support for Assange and Snowden outside the embassy, the marchers would express their implicit support for this conviction.

As they assemble on a traffic island on a sunny spring Saturday, the ambience of the event embodies the peculiar plurality that colors the comments online.[13] It is at once a site of music, of humor, of anger, and of information. The hundred or so there are gathered around a mobile sound system, listening to speeches and performances before the walk begins. A young man in a backward-facing baseball cap takes the mic and starts to freestyle. His singsong voice entertains the possibility of revolutionary cataclysm.

> Anonymous Anonymous . . . Now it's our time . . . London town, London town, we bringing it down.

The next moment, the air fills with the heavy scent of cannabis, which he acknowledges with a theatrical sniff. On the margins of the crowd, a middle-aged man in a clown wig amuses himself by honking a bicycle horn at passing cars, while a small group at the back yell out that they should all start "twerking" for their own amusement.[14] There are many Guy Fawkes masks on display, most of which face skyward. One young man, however, has his mask pulled firmly down over his face and holds up a square of black card that vents his frustration in white writing. "If you want us to listen you must listen to US!" it reads, addressing unnamed but powerful voices. Near him stands Paul, a forty-something lorry driver who has traveled fifty miles from the coast to be there. He talks about intricate features of global capitalism, such as the African debt burden and how central banks enable housing bubbles. His main adversary, though, is the mainstream media, which fail to report these mechanisms. "You won't hear the BBC questioning the stock markets," he says with surety.

Before long they set off with a resounding cheer. Along Pall Mall, onto Piccadilly, then opening out around the grand curve of Hyde Park Corner, and down toward the gleaming arcades of Knightsbridge. There are a few hundred people now bringing the traffic to a standstill. Under the warmth of the new sun and the gaze of curious onlookers, the atmosphere becomes more straightforwardly ebullient. A Brazilian lady in a raised mask and parka jacket, half walks, half dances down the emptied road. She discovered Anonymous on Facebook.

> I don't watch TV any more. My channel is Facebook. Why do I need to watch TV? We are important. We are celebrities you know?. . . Okay I'm not going to lie, I grew up listening to music, watching TV, watching movies, but one day you just woke up and you realize, whoa!

Her speech takes a more solemn tone when she recalls the changing circumstances that brought her here. After moving to the UK, she worked for nearly a decade as a carer. But while her workload increased, her pay was cut by almost half. "I just got fed up with the system, where you care for people but you're not important." Still, the past is not going to blight her positive mood.

> Look on my Facebook, at all the protests I've been to. And then you see this group, Anonymous. Oh my God! Go to Facebook and then type Anonymous. Everything will come up.

The centrality of the medium to her own presence at the event is later reinforced when she holds up a message that had been associated with Anonymous for a number of years. "Do not be alarmed," it tells the reader, "We are from the internet."

Finally the marchers swing a left onto Hans Crescent, to the tinkling of coins dropped into the case of a street musician as they pass, and enter the shady space outside the embassy. There is a moment of heightened tension when they break through the small police pen that was intended by the latter to contain them, but after that a curious lull descends. Wait a minute and you would miss it. But in that minute, just as with the arrival outside parliament at the Million Mask March, people gaze up toward the embassy as if expecting something, as if some dramatic event were about to occur. When nothing does, their previous buoyancy returns, and their attention atomizes into leisure activities. Drums are banged, horns are blown, and fingers start busying themselves with papers and rolling tobacco. A large white rectangular banner rises above their heads in the middle of the throng. It juxtaposes political concerns about corruption and austerity, with vernacular theories about the use of airborne chemicals to control the population (see **fig. 7**).[15]

The event is being administered by a tall and sturdily built White man in his late forties, with clipped hair and a square jaw. He has been a prominent figure throughout the day: in turns rallying, amusing, and instructing the crowd

FIGURE 7. Banner raised outside the Ecuadorian Embassy in London (photo by author, anonymized by Benjamin Elwyn)

through a battery-powered megaphone.[16] When Assange fails to appear at the balcony of the Embassy, it seems appropriate that he reads out the statement Assange has sent in absentia. It is general and rather short but broadly reaffirms the political importance of staging such an event. Afterward he tells the crowd that they had done what they set out to do, calling for three cheers for Julian Assange and Edward Snowden, before they set off in the direction of Downing Street. Yet once again, when they arrive outside the wrought-iron gates of the prime minister's residence, there is another curious pause. Having galloped down Whitehall, and veritably jogged straight toward it, there is a sudden uncertainty about what to do having reached this charged destination. People become strangely quiet. Again cigarettes and music fill the void, while a helicopter hums overhead. The administrator delivers his final message of the day.

> We hear your lies. We know your hypocrisy. We feel your authority. Today we march. United as one. Against worldwide government corruption.

"Knowledge Is Power, and the 99% Are Waking Up"

When Anons describe their newly awakened state, it is a condition of knowing something. In this final speech, it is not that he *thinks* the prime minister is guilty of saying one thing and doing another, he *knows* it. While adherents to earlier individuations of Anonymous are committed to the project of liberating hidden information, those in AnonUK are more likely to refer to this explicitly as "knowledge." In this subtle but significant shift, what is being accessed changes from a value-free and democratic entity that heightens the capacity for reasoned judgement, to one which is value-laden and hierarchical, and ultimately, beyond dispute. Anons talk of "finding," "gaining," and "getting" knowledge, which leads them to the elevated condition of "having" it—as if it were a precious object the possession of which makes them powerful. This elevation distinguishes them from the You who do not possess this knowledge; while placing them on an equal plane to the They, who, as a consequence, want to prevent them from getting it. Indeed, being in a condition of knowing is so vital to an identification with AnonUK, that one defines it expressly in these terms as, "a unified bunch of people who know there are things wrong."

Ostensibly, the knowledge they cultivate is through "research." For Anons, research is a solitary inquiry into a given subject, accomplished by entering terms into Internet search engines and following the hyperlinks. It may take place over the course of a few hours, but can equally go on for several months or even years. When describing these endeavors, they frequently deploy a language of exploration.

> Watching this presentation on YouTube, it just took me on a massive trip, to read and research all sorts of crazy stuff.

At the same time, Anons are adamant about the rigor of their research practice, which is achieved by tempering these open-ended explorations with methods for triangulating the information that they find.[17]

> What are you going to believe from reading one story? You have to go back and look at that university and try and find their pictures in the gallery, to find out if they actually are a professor, if they are a scientist, if they are an astrologist. Also, find what other stories they're linked to, research it all: a synopsis, a piece, a write up, a blog. There'll be some traces of them somewhere. Everything has a trace behind it. If you can follow that trace, you'll get your target.

Their conception of research overlaps with its academic corollary.[18] Precisely because of these imagined symmetries, however, it becomes increasingly important to draw out the differences. For academic researchers, any individual inquiry into a subject would modally require public legitimation by representatives of an epistemic community in the form of peer review before it could enter the category of knowledge. Moreover, academic research is generally driven by a specific research question rather than a broad curiosity, which is then answered or adapted through a variety of established data-gathering techniques. The Internet search engine is a vital resource, but these techniques will generally transcend the use of the Internet, such as carrying out interviews or using historical archives. As adumbrated in chapter 2, Anons themselves do not see a paradigmatic distinction between their own undertakings and those of academics. Indeed, as the above account suggests, their triangulation of sources may even involve reading academic papers online. For this reason, many Anons see my own role as a researcher to be fundamentally aligned with theirs, a sameness which can override other kinds of sociological differences.

Although Anons talk with enthusiasm about "learning" through these methods, they rarely talk about teaching or being taught, and if they do it is in a context of distrust. Because They want to prevent them from accessing true knowledge, teaching is something They do in order to conceal or mislead, and thereby maintain the status quo. In this vein, Anons might recall disputes they had with their former teachers, or reflect on the redundancy of what they learned at school.

> When I grew up in school, I don't remember being once taught how to plant a seed. They don't want you knowing how to be self-sufficient. If you want something, go to the supermarket.

This distrust can extend beyond a reassessment of their own biography, and percolate into ongoing discussions with their children or grandchildren, whom they encourage to question the epistemic authority of their teachers. One remarks with a degree of satisfaction that the substance of these discussions is now affecting the tutelage of his twelve-year-old daughter.

> I chat to her about anything, anything she wants to know about. And she's starting to question things now. She's hearing stuff that she is told in school and she's like, "I know that's not right dad, I know that's not right."

For one mother, what is perceived as the dangerous perfidy of secondary education was avoided altogether by home-schooling her three children from the age of ten.

Anons sometimes report their own encounter with the formal education system in unfavorable terms. Besides not being taught things considered useful, Anons describe not enjoying school more generally: whether because they were bullied by their classmates, or experienced a lack of support from would-be mentors. One man with mathematical skill now realized through computing, recounts how instead of being nurtured by his teachers, he was made to feel deviant and odd, and given a mental health diagnosis that produced a sense of alienation from a very young age. In a more extreme case, a sexagenarian describes being subjected to corporal punishment by his teachers and excluded from school activities on account of his poverty. For these reasons as well as others, these encounters tend to be shorter than the modal length.[19] In fact as a demographic, the extent of their formal education is one of the most homogeneous characteristics across all twelve key participants. Some left school as early as fourteen, while just five stayed in school until eighteen, and of these, two went on to university, one directly. Consistent with other studies of people who share vernacular theories of the world, Anons are less likely to have had the maximum exposure to the normative institutions for knowledge transference: namely, school and university.[20]

Because, for Anons, search engines offer a way to learn without being taught, they provide access to a knowledge that is uncorrupted by Them. Knowledge cultivated online is referred to unfalteringly as "truth." One woman recalls a conversation with her son, in which he queried the accuracy of some of the stories she posted on Facebook.

> My son yesterday went to me, "Mum," he said, "You put these things up but some of them might not be true." I said, "Do you actually think I'm going to put things up that aren't true? Do you not think I'm going to research this?"

As the Internet is the primary mediator of truth, conversations with Anons are peppered with injunctions to "look up" or "Google" certain persons or topics, to corroborate or supplement particular factual claims. These injunctions become palpably more frequent when I press them for further details on the project of depopulation and its authors. I am told not to "take their word for it" because the proof is there on YouTube. Dialogues such as these reveal something more about the peculiar character of the truth online. Not only is it necessarily hidden (and therefore has to be uncovered through the practice of research), but it always pertains in some way to the great societal conflict that is at the heart of their worldview.

Truth is also one side of a conceptual binary, which positions information deriving from other forms of mediation as false. Akin to their distrust of teaching as a reliable source of knowledge, information disseminated through corporate newspapers and television channels is understood to enact a conscious project of distracting or concealing the truth from mass society. They consider news disseminated through these media as at best irrelevant to what is really "going on," and at worst, propaganda to "numb" the population. If they do watch television, Anons sometimes stress, this is carefully selected documentaries rather than news or entertainment programs.[21] Although these forms of corporate mediation are normally represented as a monolith—often abbreviated to "the mainstream media"—Anons reserve a particular ire for the British public service broadcaster, the BBC. Waking up commonly involves turning away from the BBC as a news source, and in some cases boycotting the license fee through which it is funded.[22] The BBC's physical headquarters in Central London, Broadcasting House, is also the site of several Anonymous demonstrations, over and above the annual Million Mask March.

On the one-year anniversary of the march to the Ecuadorian Embassy, another is scheduled to go to Broadcasting House.[23] This time, instead of Julian Assange, another exposer of secrets is given center stage—a diminutive man with a Welsh accent and a smart navy suit displaying the golden livery of the Royal Air Force. Although this man does not self-identify as Anon, the substance of his intervention overlaps sufficiently with the values of AnonUK to allow for a collaboration in this instance. His story begins, he tells the small crowd in Trafalgar Square, when his partner lost her life savings in a fraudulent investment fund. Since that time fourteen years before, he has undertaken a long period of research which has revealed the existence of massive money-laundering schemes operating out of fake addresses in North London—claims he invites the audience to verify by typing in those addresses as search terms. The overarching purpose of the march is to deliver this information to the BBC through the front entrance of Broadcasting House, which he subsequently does as Anons cheer him in and out of the building. His hope, he announces on exiting, is that the BBC will conduct their own investigation, and those guilty of financial fraud will finally be called to account.

This episode is one of many which suggest more ambivalence in Anons' attitude to knowledge than some of their Manichean statements would suggest. If the BBC are repeatedly misinforming the population, why take the time to applaud the delivery of information to them in the hope that they will? Similarly, the frequent assemblies they stage outside Broadcasting House seem paradoxically to contain a hidden kernel of hope within them, that the broadcaster might finally help to deliver the moral reckoning that they seek.[24] Furthermore, uncertainties and doubts erupt every once in a while, inside the interstices of bold claims. Some acknowledge quietly that all the information online can be confusing and, "You don't know who's misinforming you." What all Anons are steadfast about, however—a position echoed in the Welshman's tale—is the truth-value of the experiences that instigate the process of waking up. The subjective foundations of what they know return us to the ritual sequences of traditional initiation.

Initiatory Knowledge

The liminal period is a time of intensive learning for the neophyte. What exactly is learned in this transitional stage can take a variety of forms. Some of this has to do with the transmission of skills necessary for the reproduction of ritual life. Initiands may learn how to play ceremonial instruments, or how to make their masks and costumes.[25] They might learn a special dance, or how to recite a song or incantation.[26] These forms of embodied learning can be supplemented, or even substituted, by apprenticeship into more theoretical knowledge. As adumbrated in chapter 2, liminality is designed to be a "stage of reflection," in which incumbents are invited to think deeply about the essential operations of the world.[27] Audrey Richards, in her study of the Bemba female puberty rite (a key source for

Turner's essay on liminality), relays how elders stress that the purpose of the rite is to "teach and teach and teach" the girls to prepare them for marriage, using the same word, *ukufunda*, that they use for European-style education.[28] She also observes that in practice, the skills required by married women are not learned during the ceremony, but for many years before and afterward. Instead, what the girls actually learn during their *Chisungu* are ways of thinking and talking about these activities that are unspoken in everyday life. The girls are taught clandestine word-terms, hidden meanings, and otherwise unuttered obligations that befall them. In other words, much of the deepest and most profound forms of knowledge acquired in the course of these rituals are, by definition, secret.

This commitment to secrecy is more or less important depending on the initiating community. Oaths and vows of silence may be taken as part of the rite, in order to strictly delimit the circulation of the knowledge being transmitted.[29] In such cases, initiands are compelled to keep secret what they learn from the uninitiated under the threat of powerful sanctions, sometimes even to the point of death.[30] Oaths of silence are particularly critical in the rites of secret societies. In his influential essay on secrecy, Georg Simmel argues that it is their shared preservation of secret knowledge that produces the "apartness" of these societies, forming, in essence, the substance of their collective bond.[31] To whatever degree the things that are learned in initiation rites are enveloped within promises to keep them secret, it must furthermore be added that there is always a hierarchical dimension to this apartness—an assertion of superiority that stems from knowing something that others do not. This assertion may even be folded into the language used to describe initiated persons. In the Congo, the uninitiated are referred to as *vanga*, "unenlightened," while the initiated are *nganga*, "the knowing ones."[32]

In traditional initiation, ritual subjects are deliberately placed in the vulnerable condition of symbolic death, to render them receptive to these epistemic shifts. What is fully recognized in these procedures is that social knowledge is most effectively won when underpinned by a subjective knowledge oriented around the body and its sensations. This recognition is particularly vivid in Erving Goffman's examination of the entry procedures to what he calls "total institutions"—social settings with hard boundaries between inside and outside, where members work, sleep, and play together, under whose definition can fall anything from prisons to boarding schools.[33] Goffman documents the ways in which new members are systematically "mortified," their previous sense of themselves strategically disassembled through a number of depersonalizing methods, before they begin to learn the new rules and relationships that govern the institution they have joined.[34] In each case, these subjective physical and psychological sensations are themselves a kind of learning onto which abstract knowledge is imprinted. Jean La Fontaine, in her comparative study of initiation, develops this argument most extensively.[35] Following Richards's meta-analysis of the unstated purposes of these rites, La Fontaine insists that in their ritual obligation to submit to the social and cultural authority of their initiators, what initiands really learn is the legitimacy of authority itself. As one of her own interlocutors among the Gisu tells her about the experience of being circumcised, "You cannot know what the knife is like until you feel it."[36] The implication here is that, not only is traditional initiation centered around the deliberate fusion of subjective and societal knowledge, but the social relations within which these subjective experiences take place in some form presuppose the theoretical knowledge being acquired.

The learning of theories about the basic operations of society is particularly prominent in digital initiation. But like its predecessor, this societal knowledge is written onto a subjective experience that predetermines the kind of knowledge toward which digital initiands strive. Anons talk about going online to find "something that you know already," in which what is already known is a palette of biographical experiences that have prompted the search. Because of the individual quality of these histories, and the deeply personalized character of these Internet journeys—sundering the simple hierarchies of in-person knowledge transference into countless micro-hierarchies between the producers and consumers of digital content—each Anon tells their own story about how and why the world came to be as it is. Like the lessons in cosmogony and sociogony that happen during some traditional initiations, these narratives present a mythic form of history, by asserting some primordial period of freedom before They seized the reins of power.[37] It is here we arrive at an aspect of AnonUK that draws these initiations closer to others into political cults, and away from puberty rituals into small-scale societies. Because while the purpose of the latter is nominally to apprentice the initiand into a given moral and social order so that they might reproduce it, the unorchestrated circumstances of Anons' experiences yield the individual discovery of an immoral order nested in the structures of British life. Rather than keeping this secret, and thus permitting its own reproduction, it becomes incumbent on them to expose it, injecting the impetus for change.

Exposure

Four months after the march, I sit down with one of the participants, Tank, in a dimly lit pub in the outer reaches of West London. The generous space is largely empty, and he has chosen a broad table in one corner of it, greeting me kindly from a seat facing the entrance. "I've got some things I can let you have" he opens, materializing a reinforced brown A3 envelope holding several fine color posters of graphics that he has designed and posted online over the past few months. They include promotional images for protests, alongside aphoristic reflections on societal coercion. "Now this one," he says, as he passes across a fifty-four-page science-fiction parable that he wrote the previous year, equally well presented in a transparent plastic wallet (**fig. 8**). It is a veiled account of his own experience of waking up, and acts as a springboard to an extemporization of the topic.

A vibrant three-hour discussion ensues. Tank's particular concern is with the hidden nature of contemporary power, that there are malevolent forces at work that the vast majority of people are unaware of. When he addresses this subject, his voice lowers audibly, making me wonder whether the decision to sit in a corner of the pub is to achieve a semblance of privacy.

> For me there's this huge perspective of darkness that overshadows everything, because behind the scenes there are these rich, elite, puppet masters that pull the strings, that have total control.

After ninety minutes or so he gets out his laptop. When he opens the browser, there are traces of my name on his search history, linking to the scattered details stored online, which form a random and incomplete shadow of my own biography. I point them out and we laugh about it. "I've got you!" he spurts, but in a fast way which suggests uncertainty behind the joke. The purpose of the laptop is to show me a YouTube video while he uses the facilities. Its content is consistent with the thrust of our conversation, as it uses Google Earth to identify a number of isosceles triangles that exist within the architectures of key symbolic sites across the United States, Europe, and the Middle East.[38] The video is a neat exhibition of the crucial historical role that Freemasons played in royal landscaping and urban planning across the world. For Tank, however, it is concrete evidence of his political theory. "It's a hidden message that runs through our society . . . their control over us," he says, blue eyes lighting up. For months and years afterward, Tank refers me to graphic memes and posts which turn on this theme.

Tank's background contains the curious blend of left- and right-wing politics that is discernible across different individuations of Anonymous. In his professional life he has erred on the right. After leaving school, he initially trained to be an officer in London's Metropolitan Police. This was the mid-80s, and having seen how the police were instrumentalized by Margaret Thatcher, he decided in the end not to join up. Instead he went into sales and, he says, became very money-driven—comparing himself with a laugh to the fictional financier Gordon Gekko in the film *Wall Street*.[39] Some of these choices echoed the inclinations of his father, a man he describes as property- and status-minded, and who for several decades has read the *Daily Mail*, a notorious right-wing British newspaper. Having witnessed Thatcher's conservatism at work, however, Tank became a "staunch Labour supporter," and here followed the priorities of his mother. He was particularly inspired by his mother's mother, a left-wing working-class Welsh woman, who went into household service at the age of fourteen.[40] When Tony Blair was elected with a sweeping majority for Labour in 1997, he recalls

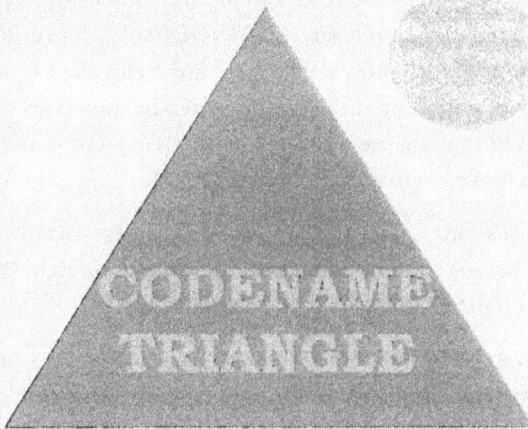

'CODENAME TRIANGLE – CONSPIRACY BEHIND TOP SECRET SPY
PROGRAMME REVEALED: EVERY INTERNET USER CONSTANTLY
WATCHED BY UNREPRESENTATIVE AUTHORITY AGENCY'

If you believed you had uncovered the biggest conspiracy of all time-
TRIANGLE, which affects how you, your family and friends, live your lives,
what would you do?

Tell no one or perhaps share this secret with a few trusted others?

Maybe you would challenge the very authority that you believed were part of
this conspiracy?

THIS ACCOUNT UNCOVERS THIS CONSPIRACY,
IS WORK OF FREEDOM,
AND IS NOT AVAILABLE ON THE INTERNET

Anthony Orwell-Huxley

1

FIGURE 8.   Science fiction parable presented to author

the sheer emotion of it, telling his mother how he wished his grandmother had still been alive to see such a day. The Iraq War crisis turned this feeling inside out. When Blair's government took Britain to war with Iraq in 2003, against the expressed wishes of the largest antiwar demonstration in British history, he was "totally gutted." Something inside him fractured. For him, it showed that there was no democracy, and he became "very disillusioned after that," his political investments falling into abeyance. The "wake-up call" arrived a decade later. In the midst of swingeing cuts to public services, the Conservative-led government spent more than £3 million staging Margaret Thatcher's funeral.[41] The audacity of it brought all the memories of Thatcherism back. He felt impelled to act. So he just started "connecting with people" on Facebook and Twitter and became integrated into AnonUK communities online. The purpose of the march was to expedite a "global spiritual enlightenment" that he believes is already happening. "Knowledge is power," he maintains.[42]

This conviction, that the mass acquisition of knowledge will drive large-scale social change, underpins the central project of unmasking, what Anons sometimes refer to as "exposure." Waking the masses would be achieved by showing them secrets.[43] In its online precursor, digital devices were the exclusive mediators for this project. As a loose association of Internet users, Anonymous harnessed itself to the unprecedented capacities of modern computing for the collection, storage, and circulation of data, spreading information that would otherwise remain concealed from public view. In AnonUK, digital dissemination is further supplemented by analog forms of information-sharing. Sheaves of printed documents, pamphlets, and compact discs containing socially critical documentaries (see **fig. 2**) are all distributed at protests to those willing to receive them. Produced by individual Anons without organizational support and at their own expense, this extra effort to share information beyond the very low opportunity and resource costs of its digital equivalent underscores the strength (and indeed hopefulness) of their belief in the possibility of mass epistemic change and its resulting social consequences. This hope extends to the ethnographer. Indeed, just as I had prepared for my meeting with Tank by arriving loaded with questions, he had also prepared in advance, bringing items for me to keep and a film that he wished me to see.

Incarnating the value of exposure, whistleblowers had heroic status in Anonymous from the start.[44] Assange, and subsequently Snowden, expedited in different ways the sudden eruption of massive secret data sets into the public sphere, and because of this became giants among affiliates, a crowd willing to make a pilgrimage to Assange's residence to honor him and call for his release. Beyond these famous figures, lesser-known whistleblowers also command respect. A few months after the march to the embassy takes place, former MI5 operative Annie Machon—who in 1996 broke her official oath of secrecy to expose the agency's monitoring of left-wing politicians among other activities—was interviewed on AnonUKRadio in a conversation about mass surveillance.[45] Even the ex-serviceman who centers the anniversary demonstration is surrounded with a reverential air. Ultimately these degrees of deference arise because Anons themselves seek to incarnate this value. In their engagements on- and offline, Anons become microwhistleblowers, endeavoring to circulate what hidden information they can among the people they meet. This again becomes clearest in their interactions with me. I am continually told to write something down or to look something up, handed keywords, in which words are keys that unlock little known digital doors.

What brings these practices together is a shared attitude that the Internet is the originating source of truth about the social world. It is not that Anons think that everything they encounter online is true—their uncertainties, disagreements, doubts, and triangulating efforts show that this is not the case—but they do believe that with enough work, they can uncover the essential truth about a given person or thing. Anons are ideologically egalitarian and individualist in their attitude to learning, maintaining that the You ultimately need to discover the truth for themselves, and yet the Internet remains the ultimate epistemic authority, and the sharing of keywords a collective submission to this authority. Truth and the Internet go hand in glove. The analog items they distribute are rarely intended to exist as stand-alone sources of information, but are normally peppered with hyperlinks and search terms to allow the corroboration, and deepening, of the knowledge they promote through subsequent online engagements. Alternative forms of knowledge transference that do not defer to the epistemic authority of the Internet—newspapers, television, books, and educational institutions—are all represented to varying degrees as false by definition. Turning off the television, in particular, is often a defining moment in accounts of waking up.

The "truth" that lies online, and its relentless exposure through research and circulation, again follows a grain set by Christian antecedents. As with other world religions that have their origins in Zoroastrianism, one of Christianity's core doctrines is that behind the visible, material world, resides an invisible, transcendent world infused with the Divine.[46] Because this hidden world is not permanently accessible in day-to-day life, it is encountered in the form of a divine revelation—sudden unexpected illuminations that transform the believer. Revelation might be thought of as the pathway between these worlds. For most of its history, the ultimate revelation was located in the text of the Bible considered to be God's own testimony, both infallibly communicated and infallibly recorded in biblical times. If the Bible were not free from error in its entirety, then the whole proposal that it is a vehicle for divine revelation would be thrown into doubt. This principle, known as biblical inerrancy, takes us very close to the relationship that Anons have to the Internet. Because it demonstrates that when communities, both real and imagined, stake their own identity around the indisputable veracity of a given medium, no challenge to this veracity can be brooked without diminishing the strength of the community built around it.[47] In contrast to scholarship, what Anons learn online can be a brittle rather than a supple kind of knowledge, unwilling to reform under the pressure of critique, and reluctant to concede its own limits. As one critical activist observes, "Anons think they're always right."

However, knowledge is persistently cloaked in the vocabulary of scientific epistemology—a language of "facts" and "research," and in which truth itself is understood in strictly scientific rather than theological terms. It is important to address the motives for this head-on. A very long set of developments underpin this tendency which extends far beyond AnonUK. As the axioms of the scientific revolution took hold in Europe from the seventeenth century onward, so too did a limited definition of legitimate knowledge. The knowledge carried in human minds and bodies was increasingly substituted for what has been called a "view from nowhere"—an entirely objective entity that existed in a separate domain.[48] This prioritizing of what the educational philosopher Rudolf Steiner calls intellectual knowledge, a purely cerebral activity rooted in materialism, over the knowledge of "soul and spirit," what I would call a subjective knowledge located in specific human beings and their engagements with the world, travels alongside the marginalizing of initiation.[49] The old European monastic and military orders, oriented around the knowledge borne by particular persons who were rendered as such by initiation, faded from power and view, replaced by new elites as subjective and objective knowledge became willfully decoupled.[50] A particularly extreme version of this logic can be found in the more recent castigation of "experts."[51] Despite being the bearers of society's intellectual knowledge, in the very act of embodying it, such persons are dismissed as partisan. In the shadow of this long trajectory, it becomes politically reasonable for Anons to wrap their knowledge in scientific language—it is the only legitimate method of verifying their insights.

As with other initiatory sequences, however, their abstract knowledge achieves its veracity through the experience of symbolic death and the social relations that summoned it into being. Like the Gisu, they know what the knife is like because they have *felt* it. The epigraph to this chapter is uttered by a White man in a baggy sweatshirt when narrating his own entry to AnonUK. After a deeply troubling episode working for a corporate pharmacy, he descended into crisis, for which from most people he received a glib kind of sympathy, paraphrased by his sarcastic usage of the idiom "there there." Among Anons, on the other hand, it is an entirely different picture. They *know* what he has gone through in a subjective sense, which is on a different plane from imagining it intellectually. The abstract knowledge that Anons carry is inflexible because it is a view, fundamentally, from somewhere. The endeavor to remake the British public sphere through the lens of these somewheres is addressed next.

SYMBOL

Alone a symbol is meaningless,
but with enough people blowing up a building can change the world.

V, in *V for Vendetta* (McTeigue 2005)

Do you know what? I'm going to give you some love. . . .

Anon

The pearly face of Big Ben fills the screen as the hands click to midnight. It starts its steady chime. "It's time," the protagonist Evey says, as the gongs roll on. Tchaikovsky's 1812 Overture—a score commemorating military victory—begins to play from loudspeakers placed all over London. Cymbals crash and the brass section reaches a glorious crescendo, as a series of explosions rip through the base of Britain's Houses of Parliament, smashing windows and firing burning missiles into the River Thames. Then the camera pans up the column of Big Ben as the explosions mount, pausing for a split second on the sight of its iconic clock, before they finally erupt out of it. Glass shatters and dials scatter, to the discordant clang of the bell being struck. The celebratory atmosphere within which this scene at the end of V for Vendetta is framed, is completed by a bloom of fireworks in pink, green, and gold, that shoot into the night sky like a bouquet of flowers.[1]

Cults are world-building social formations. They are attempts—to varying organizational degrees—by a minority to turn away from the norms and values of a majority and inaugurate a different moral-cum-cosmological order. This is self-evidently, for members, an improved order, a progression from a lower to a higher state of moral being. Cults (and this tendency continues fully into online cults) thus generally possess a Janus-faced quality, one that looks inward and outward at once. While they look inward, with effusion during the period of growth, at the new world they are ushering into being, they also look outward, with criticism, at the world they are parting themselves from. This outward gaze can be characterized by anything from a mild despair to a violent confrontation.

In AnonUK, any form of violence against the majority is symbolic, and takes place virtually. These symbolic acts begin with this spectacular immolation of Britain's Houses of Parliament, itself a rehearsal of the memory of Guy Fawkes, and the history of opposition to parliamentary hegemony he represents. Imagining the annihilation of the dominant symbol of the British state opens up a temporality for rebuilding.

This tendency to look both inward and outward can manifest itself in the contradictory application of shared symbols. Many of the most extreme examples of this phenomenon can be found in descriptions of the cults that burgeoned across the Pacific during the colonial period.[2] Broadly speaking, these were syncretic Indigenous Christian revivals that flourished in deeply asymmetrical social conditions and undertook rites that mimicked European activities to bring about a new Millennium of material bounty. In the Hauhau cult of New Zealand, for example, the rites took place around a mast salvaged from a European ship that had run aground which was adorned with flags, pendants, and the crosses of Saint Andrew and Saint George.[3] Alongside these European symbols, the ritual assemblage contained a Maori symbol, specifically a diagram of a protruding tongue, associated with generative and defiant power. In this chapter I explore parallel symbolic contradictions and syncretisms at work in AnonUK—in which they draw on existing symbolic registers, and combine them with Anonymous iconography to convey a wildly different meaning. Indeed this was presaged by all the main symbols associated with Anonymous at its inception—namely the flag, the suit, and the Guy Fawkes mask—all of which were re-creations rather than novelties as such. The Anonymous flag, for instance (see **fig. 9**), is a reworking of

FIGURE 9.   The Anonymous flag (Wikimedia Commons)

the flag of the United Nations, superimposing onto its geometric globe a headless figure in a suit.

This chapter simultaneously follows the story of an event called OpRealLove. Taking place on Valentine's Day in the city of Peterborough, the event reimagines a festival traditionally celebrating romantic love, as one oriented instead around distributing needed goods to the urban homeless. While performing this decommodification of the British public sphere, it also participates in a practice that became popular in cultic milieus from the 1970s onward—often referred to as "love bombing."[4] Love bombing is the substitution of existing intimate and familiar ties, with affection from other members, in ways that can exert a powerful appeal. In AnonUK as elsewhere, that these extraordinary displays of affect can be efficacious as a form of recruitment does not necessarily mean they are being performed instrumentally. They can also be seen as an authentic expression of the kind of affective connection to the world that Anons seek to restore.

Defacing the Dominant Symbol

In his study of Ndembu ritual, Victor Turner develops the concept of the dominant symbol. A dominant symbol is an object which represents the axiomatic values of a given society, and provides a focus for ritual life. His paradigmatic example is drawn from their female puberty rite, the *Nkang'a*. The rite is centered around the *mudyi* tree, known as the "milk tree," with a peculiar bark that emits white milky beads when it is scratched.[5] The *mudyi* is symbolically associated with breast-milk, and through this with the bond of nurturing between mother and child that binds the matrilineal society together. Beyond this, the tree also represents the primordial ancestress from which the whole society springs. It is the dominant symbol. Yet what Turner observes in practice are the conflicts and contradictions that pervade the rituals around the tree. The tree is not only the source of life and social harmony, but equally the site of social segmentation, of mother-child separation, and the symbolic death of female initiands. For Turner, these ambiguities (which often take place without being verbally acknowledged) are critical to its efficacy. It is precisely the ability of a dominant symbol to incorporate axiomatic values, as well as their contravention, that allows it to hold the society together, dissolving both into a greater unity. But what happens when these contradictions cannot be resolved?

The Houses of Parliament serves as the dominant symbol of the British state. It is the meeting house of the nation, concentrating the axiomatic value of democracy by making every citizen present through the body of their elected representative. The weekly performance of Prime Minister's questions rehearses the principle of democratic accountability, while the repudiation of the monarch's representative at the opening of Parliament signals the supremacy of popular sovereignty over kingship.[6] But here too, activities in and around the chambers betray some of the society's many contradictions. There are still hereditary peers in the House of Lords; while its sumptuous rooms may be hired out for a fee by corporate clients. Meanwhile, the area around the building is a magnet for protest, both passing demonstrations and permanent encampments.[7] The Houses of Parliament is the first, and most significant, destination of the Million Mask March. Yet when Anons appear outside Parliament, they are not seeking to interpellate its occupants, to be dissolved into its complex unity. These appearances began as an explicit reenactment of the above scene of destruction in *V for Vendetta*, although the only part of this spectacle that actually materializes are the fireworks. In their most significant annual ritual, Anons repeatedly rehearse the destruction of Britain's dominant symbol.

This scene is another form of defacement, in Michael Taussig's terms.[8] Not only does it entail the desecration of a public symbol, but of *the* dominant symbol of the society. An additional aspect of Taussig's theory becomes more obvious here. Taussig argues that when a symbol is defaced, a "strange surplus of negative energy" is released into those around it, whose strength and character corresponds to the social order it had been holding.[9] Indeed, it is in this moment of defacement, he contests, that the spiritual power of the symbol is at its most potent. This emission of energy is made fully present in the film, through Tchaikovsky's victorious overture and the joyful effusion of fireworks. Furthermore, these are subsequently followed by a wave of unmasking that ripples through those who have gathered to witness it, all of whom deliberatively remove their Guy Fawkes masks to display their own faces, illuminated by the amber glow of explosions. As Taussig stresses, this revelatory moment and its potency can only ever be temporary, before the mundane world returns to reassert itself. Because this scene happens at the very end of the film, however, this mundanity never arrives, and so this moment seeps outward into its audiences as they switch off their screens, providing a continual source of negative energy in the looping realm of computing.

It is defacement too because there is another face being removed—the face of a gigantic clock. Big Ben is an important protagonist in *V for Vendetta*, appearing throughout the film at key moments, and its final immolation is also the end-point of these recurring motifs. It is most obviously a symbol of the modern relationship between power and clock-time. E. P. Thompson famously showed, through a combination of ethnographic and archival records, how Britain structurally transitioned from the task-oriented time of agrarian feudalism, to the mechanical clock-time of industrial capitalism.[10] In the latter, clocks became vital mechanisms of labor discipline, strategically used by factory overseers and owners to govern workers. This particular clock, however, is not just of national significance as an appendage of the dominant symbol of this modern industrial state, but of international significance as a symbol of its former imperial dominion. Big Ben still receives its reading direct from the Royal Observatory at Greenwich, established as the zero hour of global timekeeping in the late nineteenth century.[11] There is also an anticolonial subtext to the film, and the first time Big Ben appears, it is beside a dialogue in which America is referred to as one of Britain's "colonies."[12] Destroying the face of Big Ben is to blast not only the enduring relationship between capitalism and clock-time, but also, more broadly, Britain as a center of imperial power. Both of these ideas reverberate in AnonUK.

Anons exhibit a suspicion of clock-time in general, positioning themselves instead inside a constant present. Consider the following statement, which draws on New Age theories of consciousness.[13]

> When you're focused on beingness, which is ever present, you don't experience a moving of time. So to talk about past and future, it's like "What are you talking about?" It's now, only the now . . . Like being here right now, where is the movement of time? You see clocks may be moving, so that's time.

This repudiation of time as a linear continuum that connects past, present, and future is also manifested in how they themselves think about these things. A common conviction in AnonUK is that the telling of history has been manipulated by Them to conceal an unpalatable truth, namely the hidden history of Their enduring dominion. This entails both a national and international rethinking of the past.

> Victorian England was impoverished and children went to work in workhouses, but we were given this cozy image you know?
>
> I *loved* history. History was my favorite subject. And I said to Jonny, I said, "History's a lie," and it really is. I used to go to Saturday morning pictures, put Cowboy and Indian films on. "Look at all the Cowboys, they're killing all the bad people." Hold on a minute! Now we know *you're* the bad people. They had every right to live there—it was their land.

For one member, the nonlinearity of time ramifies into forms of prophesy enabled by Internet search engines.

> I look into future earth history—30,000 years from now, 50,000 years from now—how it actually played out. Did the Illuminati win?

Just as their vernacular theories of the world and its tripartite social structure position all history as the history of Their dominion, in a statement such as this, the future becomes an arena in which this axial conflict continues to unfold.

Anons' dismissal of time as a linear continuum shares resemblances with what Walter Benjamin called the *Jetztzeit*—the now-time.[14] In the early twentieth century, Benjamin launched a pioneering critique of the modern concept of progress, and its construction of time as an empty space in which the relation between past, present, and future had already been decided and determined. He saw this colorless empty time (which was also the time of clocks and other mechanical timepieces) as fundamentally belonging to ruling classes, who sought to control narratives about the past in order to foreclose potentials for the future. Benjamin relays that on the first evening of skirmishes in Paris's 1830 Revolution, clocks around the city were simultaneously and independently fired on—an actual historical precedent for this scene in the film. For Benjamin, as for the Wachowskis, this assault on the clock is a critical symbolic act, which for the former explicitly and the latter implicitly pries open a now-time replete with revolutionary potential. For Victor Turner, this disassembly of quotidian time is also critical to liminality. The liminal condition, he says, is a "moment in and out of time," and it is precisely this interstitial relation to the normal time-sense that creates the space for transformation.[15]

This audiovisual defacement of Britain's dominant symbol and its iconic clock plays a direct role in the genesis of the Million Mask Marches. And yet the effect of basing an embodied event upon a virtual one is curious. For many different types of political movements—particularly those rooted in the more machismo forms of antifascism—their activism finds an important expression in the defacement or destruction of hegemonic symbols. But when Anons arrive outside key symbolic sites, they can display a strange inhibition. As described, these arrivals can be characterized by lulls and pauses, and/or a general disinclination toward substantial property damage. Unlike V, Anons do not blow up this building to change the world, and they never articulate an intention of doing so. As an online cult in which embodied and digital (and indeed fictional and nonfictional) events are seamlessly intertwined, there is no imperative to deface—because, in effect it has already happened. Their most meaningful defacements take place virtually, which appears to purge the impulse to do so physically. Instead, these imagined desecrations leave the public sphere open to being reinscribed, one of which takes place on Britain's annual festival of romance.

#OpRealLove

The city of Peterborough lies eighty-seven miles to the north of London, in the region of East Anglia. Traveling there by train from the south you pass over the area's flat fen landscape, with its distant horizons dotted with farmhouses and trees, and arable fields lined with waterlogged trenches—a reminder that the land was once reclaimed from rivers and tides.[16] The earth below is rich in a gray combustible clay; and it was the discovery of this that positioned Peterborough as a center for brick-making in the late nineteenth century, a development made possible by the arrival of the railway in 1850.[17] In the early twentieth century, its reputation as a center of industry grew. Firms were drawn by the availability of cheap labor banished from the land by automation, as well as the proximity to London, and textile, sugar beet, and machine factories moved in. It was the latter that came to define Peterborough's workforce for much of the rest of the century. Two firms—Baker Perkins founded in 1923 and Perkins Engines in 1932—established Peterborough as one of the world centers for precision engineering, producing industrial machinery, and some of the first high-powered diesel engines. These, alongside other engineering concerns, were some of the city's biggest employers until the 1970s, when during the slump they shrank drastically, and by the 1990s almost all had closed down—the Baker Perkins factory ultimately becoming what is now Peterborough prison. In the 1970s and 80s, city councilors responded to high levels of unemployment with the promotion of new white collar sectors: travel agents, life insurance, and software companies all established their headquarters there, supported by urban planning initiatives and housing estates.[18] In the twenty-first century the industrial landscape changed again, with the spread of large supermarkets and gargantuan distribution centers—introducing a new era of unskilled retail and warehouse work.[19] Platform companies like Amazon now employ thousands of workers in Peterborough, sometimes simply on a seasonal basis.[20]

Upon arrival, the architectural tableau has the uneven look of somewhere which has been aggressively modernized over successive historical periods—a subject of its strategic advantages. The station concourse is chrome, glass, and sibilantly swishing gates, while the meaty structure of the Great Northern Hotel, which arrived with the first railway, still stands opposite. As you follow the well-marked pedestrian path toward the city center, you are greeted by the kind of urban planning that is peculiar to the twentieth century—a fast-moving dual carriageway with a four-story car park peering over it. It is hard to imagine that just over a century and a half ago, Peterborough was a small rural market town built around an ancient cathedral. The unevenness endures as you make your way in. The path takes you directly onto the main thoroughfare of Cowgate, a now-trafficked road lined with a mixture of Victorian buildings with tall windows and chimneys, and others designed in the geometric shapes of the late twentieth-century office. At ground level sits a mixture of estate agents, insurance firms, and a range of pubs and restaurants. Peterborough's deep Christian past becomes clearer as the road pans out into the generous pedestrian expanse of Cathedral Square. First you pass a substantial fifteenth-century Baptist church, before moving toward the looming Gothic spires of Peterborough Cathedral.

Where the cathedral now stands was originally the site of a Roman settlement, and the Roman influence continued into the sixth century. A group of forty monks were sent by the pope to rid the local Anglo-Saxons of their paganism and bring Christianity to the region.[21] The first abbey was built here in the seventh century, and in those that followed it became a locus for monastic production, its orders engaged in making sculptures and fine illuminated manuscripts.[22] As a symbolically rich location from a very early stage of British history, the abbey and its successors were also the site of multiple destructions and defacements. The first building was destroyed by Vikings, and the second ransacked by Normans. It is the third abbey built on this site, erected in the early twelfth century, that still constitutes a part of the cathedral today. Having been one of the few abbeys to survive the dissolution of the monasteries, the third and final major symbolic assault came in the seventeenth century. Puritan iconoclasts led by Oliver Cromwell carried out irreparable damage to the building over a matter of weeks—smashing stained glass, breaking down doors, and destroying books and carvings and other artifacts considered idolatrous.[23] Indeed, on the front face of the cathedral are small noticeable voids where stone figures once stood.

But this destructive and disfiguring relationship to symbols is a world away from the event that Anons have been working toward for the past month. Because they have confined their most important acts of symbolic violence to the virtual, the atmosphere is resolutely placid. In the center of Cathedral Square sits a sandstone building that looks something between a house and a gazebo, providing shelter from the elements while remaining open on all sides. The interior has been packed with clothes and other items. Smartly suspended on half a dozen rails are row upon row of men's and women's jackets, jumpers, trousers, and T-shirts, while a mountain of unsorted bags in one corner indicates that there is much more than can be displayed. Along one perimeter are laundry bags brimming with old scarves and new socks, while on another are bowls of snacks and bottles of water, and an array of handmade and professionally printed signs. The iconography of love abounds. Scarlet hearts embroider messages reading, "The answer: LOVE" and "Bringing Community Spirit Home." The dozen or so men and women staging the event all appear to be over the age of thirty, and are all White with the exception of one Black participant. They move around the space—sorting, setting, tidying—with a quiet animation and a purposeful, even officious, air. Over their winter coats they wear high-visibility vests that have been specially printed with the name of the event. From a distance you might think that they were contracted employees, or working for the city council. Yet the Guy Fawkes masks which pepper the scene indicate, as ever, that all is not what it seems.

The event is known as OpRealLove and these goods are intended for the homeless denizens of Peterborough.[24] The prefix op- references its heritage in Anonymous's online operations, while the idiom of the real announces its super-imposition onto aspects of Valentine's Day that are considered false. One of the key administrators tells the local newspaper beforehand that "the real essence of love" has become confused with "diamonds and chocolate." This sentiment is echoed on the day.

> People think "Let's show love. Let's spend a fortune on goods for our loved ones," and their justification is that the whole world is doing it. It's all manufactured and it's all about sales, really, for me. They just make the generic cards. And I am out of that loop now. I was in it, admittedly, I was in the loop about Valentine's Day. I bought into it.

The implication in these statements is that there is an authentic spiritual element to the festival that has been co-opted. In a vernacular elaboration of what Marx named the commodity fetish—when objects appear to be charged with a life that is only the charismatic effect of their exchange-value—the second speaker distances himself from it.[25] In this act of distancing from one set of relationships that normally organize the ritual, it allows those present to reweave another. In this narrative of moral change, he adumbrates a transition from one nexus (or "loop") to the next: in which the former is characterized by the blending of romantic intimacy and the commodity-form, and the latter, by the care of homeless people and the gift—or as I elaborate in the final chapter, sacrifice.

OpRealLove is a place where everything is free. I learn quickly that money is taboo when I am approached in the morning by Bez—a lean man with hollow cheeks and large solemn eyes—to see whether I would like a hot drink. He is clasping a piece of white paper on which a list of other requests have already been penned, and when I ask for a hot chocolate his expression visibly softens, before he conscientiously writes down the order. Can I give you some money? I offer, anticipating the reply. "No, it's my treat" he says definitively, before moving off to ask others. This tiny exchange is just the start. Everything in sight has been donated, and not just from generous individuals. A mountaineering shop on the high street has been persuaded to contribute £500 of winter clothing, and there are other smaller donations from supermarkets and clothing chains. Plucked from the end of long chains of production and distribution, the ultimate destiny of these goods as items for purchase is subverted, moments before its final realization. Although OpRealLove is a place where everything is free, it should be observed that there are only a handful of homeless people there to benefit.

The Peterborough Anons have counted a total of forty-eight homeless residents, and before the event had made a concerted effort to publicize it, placing flyers in shelters, and perambulating the city with heart-shaped balloons to encourage them over. Yet only a handful appear. This seems to matter surprisingly little, as it is a spirit of sacrifice that unites those present, not the more transactional mode of donor and recipient.

After the bustle of setting up the space subsides, large gaps of inactivity yawn that are filled with conversation and play. A wiry energetic man called Alex relays a tale of separation that spans several decades. He began suffering from heart problems at the age of fourteen and was eventually forced to stop work in his twenties, spending seven years on disability benefit. By the time he was twenty-seven, Alex's health had dramatically declined and he received an emergency heart transplant. Since that operation, he feels like he has had seventeen years "extra," and the specter of death permeates his account. He returns to the subject of his father's passing three years ago (a former miner in England's Midlands) repeatedly, reflecting on the profound effect that it had on him. The final shift came last year when he was working in a large logistics center nearby, and his health suddenly started to deteriorate again, compelling him out of work and back onto state support. Finally, he was told, he would need another heart transplant. He responded to this news with a dramatic exercise in material detachment—jettisoning books, CDs, his stereo, virtually everything he owned. "It's just stuff," he reasons, "So I got rid of it. It's dead." All that remained was a laptop, a sofa, and a small set of clothes. Today he is dressed head to toe in his normal color—black—in a visored hat, waterproof jacket, combat trousers, and trainers.

Alex is remarkably upbeat, and he explains all this without a tincture of self-pity, casually removing a pill-box from his pocket and swallowing its contents as he speaks. After the pit of his crisis came a process of recovery attended by his new association with AnonUK. It began when a neighbor encouraged him to watch a presentation on YouTube. He duly did and an immersive period of online research followed, prompting him to assemble friends on Facebook who identified as Anon. At the time he was spending his days in and out of hospital, physically incapacitated, and "tearing his hair out." Anonymous's message about the universality of suffering and its practical solution—supporting the homeless—hit home.

> Because we *all* suffer, we *all* have tough times, and we can either curl up in a ball and cry and *die,* or we can get up and be strong and help people in similar situations.

Engaging in activities like OpRealLove gives him a "feeling of worth." After his exercise in detachment, Alex undertook a form of refacement.[26] He used the money he made from selling his possessions to embroider the backs of his hands with colorful tattoos, and on the upper left-hand side of his jacket he has painted a white loveheart. It looks like a small memorial to the heart that he lost. Like many Anons, Alex pulls his mask down when posing for a photo. Echoing his own grunge aesthetic, its plastic lip has been pierced by a tiny silver ring.

Another man with a mask on his head does not share this sense of resolution. Garvey is a tall Black man in his mid-forties, with a soft and slow-moving, but steady, demeanor, who calls himself the BFG—the Big Friendly Giant of Roald Dahl's children's story. Garvey was charged in the early 1990s with a murder, that historic evidence revealed since the case came to court, suggests he did not commit. He spent twelve years in prison, and after that was released on life license. He now has a weekly day job in catering, but can still be recalled by the police at any moment. Since Garvey was released, he has campaigned relentlessly for his case to be reviewed, and to receive a full pardon—a campaign which has featured on the mainstream media and been discussed by his sitting MP. Garvey is not a quintessential Anon. He makes no reference to the discourse of consciousness, nor to the transformative effects of the digital. But he is an example of the broad range of political projects that AnonUK is able to encompass at its summit. Like Alex, it is the leap from the subjective to the social that makes OpRealLove meaningful. "Too many injustices have been happening," he says, "and it's like the injustice that happened to me."[27]

Around lunchtime OpRealLove draws to a natural close. The vast surplus of untransacted goods are restored to their bags and boxes, and piled into a white transit van that has pulled onto the square. I jump onto one of the passenger seats beside Lily—an influential Anon with dyed red hair and beguiling, secretive eyes—and we make our way to her house where it will all be stored. She lives in a two-story dwelling on one of Peterborough's many housing developments, with a handful of different vegetables sprouting in the front garden. Once the others have arrived, we naturally form a human chain, people positioning themselves at intervals inside and out (with those in the hallway downstairs winding their way around a large motorbike) to carry the bags upstairs. They are ultimately flung by Bez into one corner of Lily's bedroom, a space painted uniformly black, and soon begin to form a mountain taller than human height. Somehow, Lily observes, it is even taller than it was when it was all removed that morning. She is unfazed though by its dominion over her domestic realm, "As long as I can get into bed," she says nonchalantly. It is taxing work, and once finished we rehydrate with pints of squash before clambering into her car. On the back are several bumper stickers, one of which contains a message somewhere between a joke, a gesture of love, and a declaration of war—"MY KARMA RAN OVER YOUR DOGMA."

Back beside the sandstone structure of Cathedral Square, the day enters its more expressly political second stage. The ten or so that remain, undertake a two-and-a-half mile pilgrimage to Peterborough's main NHS hospital, to protest the encroaching privatization of the National Health Service. It sets off with an initial flourish of chants and cheers—"Whose NHS? Our NHS!"—which subside rapidly into an awkward silence that is brushed away with self-effacing humor. "We're going to need it after this!" a middle-aged man adds to affirmative chortles. With a new range of printed placards now propped on their neon shoulders, they again inhabit existing symbolic registers. You would be forgiven for thinking, at a distance, that they were working on behalf on of an estate agent, as they carry large FOR SALE signs that reproduce the aesthetics of those planted outside newly commodified buildings. It is only the less visible subtitle that contradicts this first impression, the NHS is actually "not for sale," it asserts (see **fig. 10**). Before long we are tracking the busy dual-carriageway beside the station, and the air fills with the roaring crescendo of passing cars. A number of drivers honk their horns in support, and the Anons rarely fail to cheer or wiggle their placards in reply. The mood is high. Bez, who has been live-streaming the event by walking backwards for much of the way, exclaims to the group, "We've got seven live viewers!" to more jubilation.

We arrive at the hospital gates significantly later than advertised on Facebook. Banishing clock-time has the practical effect of banishing those who attend to it, as the police officers and security guards who had been stationed there to supervise the protest, had long since departed. This removes the possibility for a confrontation with state power that characterizes some other AnonUK events, and instead they smile and wave at the nurses as they drive in and out. This particular arrival is distinct from others too, because the building embodies the Left hand of the British state, its capacity to care for its citizens on mass scale.[28] While the Million Mask March re-enacts the spectacular immolation of the Houses of Parliament, considered, in line with the film, as the ultimate symbol of the state's Right hand—its capacity to coerce—they venerate the NHS and its institutions in full. For them there is no contradiction. There is also an observable symmetry between their homage to the hospital, and their previous activity in the square. Both OpRealLove and Peterborough Hospital—albeit with very different resources at their disposal—share the same raison d'être, to provide care free of charge to those who need it. Indeed even in this era of austerity, the NHS remains a state institution before which, theoretically at least, Britain's homeless citizens remain entirely visible.

The nominal reason for their presence is to protest the quiet subversion of this raison d'être by profit-making actors. The marchers are not expansive on the details, but at the time of the event the amount of NHS revenue spent on private

FIGURE 10. Photograph of signage at OpRealLove Peterborough (photo by author, anonymized by Benjamin Elwyn)

provision has risen to over ten per cent of its total.[29] But beneath this, it also allows them to impose themselves on the space, reinscribing it in their image. Several Anons spend some time writing the word ANONYMOUS in wet mud on the tarmac path with the balls of their feet, before standing for photos behind it, arms crossed, masks down. The various placards they have brought with them are attached to objects around the scene with duct tape, one of which articulates this sense of agency they have pulled into themselves.

> I always wondered why somebody wasn't doing something about this. Then I realized I am somebody!

Alex, normally cautious about straining his new heart, remarks that this is the furthest he has walked in months. In the ease of completion, we spend a couple of hours there. They pass around cigarettes and spliffs, before the grey sky fades and the winter cold creeps beneath our coats. Some depart by bus, while the rest of us bundle into two cars that have pulled up on the verge. One person asks how exactly we are going to fit ten people onto eight seats. "Don't worry," Lily says, a gleam in her green and yellow-flecked eyes, "We're Anonymous." It is a joke, and we laugh. Yet in the manner of most jokes there is an ambient truth lying just behind it.

Calendrical Time and Transformation

OpRealLove is one of several re-makings of the British ritual calendar. If you consider the events in **table 2** that are administered directly by Anonymous, besides those that respond to immediate political developments, it is exceptional for these *not* to fall on or around a symbolically significant day. To Valentine's Day we can add Bonfire Night, Christmas, and the August Bank Holiday, as well as lesser-known activist anniversaries marking Martin Luther King, and Julian Assange respectively. Others simply fall on the weekend. From a pragmatic perspective this would suggest that the modal working week is still significant for participants, but it also has the effect of re-imagining how the weekend is widely used in Britain, namely for the consumption of goods and leisure. One event, titled "Elf not Wealth," falls on the Saturday before Christmas, a day on which British high streets are reliably thronged with shoppers. As an alternative, Anons gather in Trafalgar Square to stage their opposition to "austerity measures and corporate greed over the festive period," by distributing presents to the homeless around central London, a subversion comparable to OpRealLove. Another of their events falls on the conjuncture of the weekend with the first day of the month, calling itself "Revolution Day." Revolution is a word Anons use frequently to describe these activities, but their consistent repudiation of linear time in favor of the cyclical time of weeks, months, and years, brings its meaning much closer to the pre-modern association of the word—return.

In the same essay on the concept of history, Walter Benjamin explores calendrical time as a different form of consciousness to the time of clocks and progress. Holidays and festivals are, he offers, days of remembrance, which through the act of repetition look backwards to their previous incarnations, rather than forwards into the future's uncertainty.[30] The difference between these two kinds of consciousness is more fully articulated by Mircea Eliade.[31] Eliade sets up a structural antithesis between the linear and developmental concept of time that emerged in Europe after the French Revolution, and a pre-modern and "primitive" idea of time as cyclical, in which each cycle sought to restore a mythic archetype and with it to instigate a new beginning. The first concept of time reifies the idea of historical change, yet Eliade makes the somewhat counter-intuitive claim that cyclical time is in fact much "more creative," as it allows ritual subjects to fundamentally regenerate themselves at the birth of each new cycle.[32] Calendar rituals like Valentine's Day produce a spiritual purification, annually expelling sins and demons represented, in this instance, in secular terms. The reader will observe a slippage beyond the limits of his analysis. For me, Eliade erred in seeing this second concept of time as "archaic," and having nothing to do with contemporary

TABLE 2. All chronicled fieldwork events

DAY	DATE	START TIME	NAME	ORGANIZED BY	PLACE	NOTABLE DAY
Saturday	18 January 2014	1:00 pm	Next March for Freedom UK	Anon Facebook Admin	Trafalgar to Parliament Square	Weekend
Wednesday	19 February 2014	10:00 am	ATOS National DEMO	Facebook Admin	Atos HQ, Triton Square, London	No
Saturday	1 March 2014	1:00 pm	Worldwide March Against Government Corruption	Anon Facebook Admin	Trafalgar Square to Ecuadorian Embassy	Weekend
Friday	4 April 2014	12:00 pm	Worldwide Wave of Action	Anon Facebook Admin	Trafalgar Square to St. Paul's Cathedral	Anniversary of assassination of Martin Luther King
Saturday	19 April 2014	1:00 pm	Occupy the BBC: March Against Mainstream Media	Anon Facebook Admin	BBC Broadcasting House, Portland Place, London	Weekend
Thursday	1 May 2014	3:30 pm	Occupy Wonga	Occupy London	Trafalgar Square to Wonga HQ, Hampstead Road	May Day
Thursday	22 May 2014	8:00 pm	OpFeedTheHomeless	Anon Facebook Admin	The Haymarket, Norwich	No
Thursday	19 June 2014	2:00 pm	Solidarity Vigil for Julian Assange	Anon Facebook Admin	Ecuadorian Embassy	Anniversary of two years at the embassy for Julian Assange
Saturday	21 June 2014	1:00 pm	No More Austerity. Demand an Alternative	The People's Assembly	Portland Place to Parliament Square	Summer Solstice
Saturday	19 July 2014	12:00 pm	National Demonstration for Gaza. Free Palestine!	Coalition of Pro-Palestinian Organizations	Whitehall to Israeli Embassy	Weekend
Thursday	24 July 2014	10:00 am	London M25 Blockade: Battle for Britain's Children	Misc Facebook Admin	South Mimms to Thurrock Services along M25	No
Friday–Sunday	29–31 August 2014	9:00 am	Mass GCHQ Protest 2014	Anon Facebook Admin	GCHQ, Hubble Road, Cheltenham	Bank holiday weekend

Day	Date	Time	Event	Organizer	Location	Notes
Friday	26 September 2014	7:00 pm	Occupy Whitehall and Parliament Square	Anon Facebook Admin	Parliament Square	Parliamentary vote on Syrian airstrikes
Friday	17 October 2014	5:00 pm	Occupy Democracy	Occupy London	Parliament Square	Day before TUC Demonstration
Saturday	18 October 2014	1:00 pm	Britain needs a pay rise	Trade Union Congress	Trafalgar Square to Hyde Park	Weekend
Saturday	1 November 2014	12:00 pm	Anonymous Paperstorm	Anon Facebook Admin	Trafalgar Square to Embankment Bridge and back	Saturday before MMM
Wednesday	5 November 2014	6:00 pm	Million Mask March	Anon Facebook Admin	Trafalgar Square to sites in central London	Bonfire night
Saturday	13 December 2014	1:00 pm	OpSafeWinter Nottingham	Anon Facebook Admin	Old Market Square, Nottingham	Before Christmas
Saturday	20 December 2014	1:00 pm	Elf not Wealth	Anon Facebook Admin	Trafalgar Square	Before Christmas
Friday	23 January 2015	10:00 am	Eviction—The Fraud of the Bank	Tom Crawford	Carlton, Nottingham	Day of a home eviction
Saturday	14 February 2015	10:00 am	OpRealLove	Anon Facebook Admin	Cathedral Square, Peterborough	Valentine's Day
Sunday	1 March 2015	12:30 pm	Revolution Day	Anon Facebook Admin	Trafalgar Square to BBC to Parliament Square	First of the month, weekend
Wednesday	15 April 2015	2:00 pm	March for the Homeless	Anon Facebook Admin	Piccadilly Gardens to Albert Square, Manchester	No
Saturday	2 May 2015	12:00 pm	Global Cannabis Day	Feed the Birds & London Cannabis Club	Downing Street, central London, Parliament Square	Global Cannabis Day
Saturday	9 May 9 2015	12:30 pm	Anons against the Torys [sic]	Anon Facebook Admin	Downing Street, central London, Parliament Square	Saturday after the UK general election
Thursday	5 November 2015	1:00 pm	Million Mask March	Anon Facebook Admin	Trafalgar Square to sites in central London	Bonfire night
Saturday	5 November 2016	6:00 pm	Million Mask March	Anon Facebook Admin	Trafalgar Square to sites in central London	Bonfire night

Europe.³³ Just the history of Guy Fawkes Day alone illustrates the perdurance of practices he would think of as primitive, in the very foundations of the British polity.

By far the most significant, and the only truly repetitious, date on the Anonymous calendar is 5 November. The political origins of this festival are rooted in an act passed by the English parliament in 1606, when it was agreed that the day would become one of "perpetual remembrance" for the delivery of King James I from his would-be assassins.³⁴ At the time, the monarch's narrow escape was considered to be proof of divine intercession, and the decision to memorialize it a means of reinforcing this interpretation of divine support for the ascent of the Protestant establishment after the Reformation. What was being remembered was Catholic failure. The new Protestant regime embodied by James I, was also a centralizing and territorially expanding one. King James I of England was simultaneously King James VI of Scotland, and the binding of the two monarchies laid the constitutional foundations for the Act of Union that merged the two countries a century later. In possible recognition of the significance that this day already had for English Protestantism, it was also on 5 November that staunch Protestant William of Orange landed in England to assume the crown. Hence from its earliest years, this date was inextricably linked to the emerging consciousness of the British nation, and the beginnings of its imperial expansion. David Cressy goes as far as to argue, that in the absence of any other day of independence or revolution, November 5th is in fact the closest thing that Britain has to a national anniversary.³⁵

But the act of remembering, of restoring an original archetype can be, as Eliade observes, a fundamentally creative process. Through repetition, something else might arise, and with November 5th this happened almost instantly. As adumbrated above, alongside the official celebrations, a rich counter-hegemonic tradition soon developed, and in the nineteenth-century somewhat surpassed its sober sibling. The escalating rowdiness around Guy Fawkes night led British parliamentarians to pass a subsequent Act in 1859, removing 5 November from the official Church of England calendar. The Million Mask March is a continuation of this counter-hegemonic tradition and serves, to some degree, as a mirror of the British polity that this date has long commemorated. In contrast to nationalism, it expresses everything but: Anons hold up Union Jacks on which subversive messages are scrawled (see **fig. 3**), while the flags of the counties of Cornwall, Derbyshire, and Yorkshire can all be seen substituting national with sub-national identities. In 2014, the march takes place just months before the Scottish referendum on independence, and in a historical counterpoint to King James's union of the two kingdoms, independence activists gather in Glasgow and Edinburgh, wearing Guy Fawkes masks designed in the colors and shapes of the Saltire.

Turner emphasizes the new thought and custom that flourish when structure loses its legitimacy. The disintegration of established positions and the connections between them, allows new ways of being and relating to be forged. As elsewhere in Anonymous, what is being culturally created grows out of a Manichean dichotomy between the true and false, between something that is living and "real," and something moribund encrusted over it, that is in the act of being removed. OpRealLove is expressive of this broader dynamic of unmasking that defines Anonymous's symbolic work. Mircea Eliade's hard evolutionism inhibits the application of his thought to this context, yet what he lucidly recognizes, is the revitalizing effect of this kind of cyclical time that is being performed through this re-imagination of the British calendar. Calendrical rituals—particularly the fire rituals of which Britain's Bonfire Night is the contemporary inheritor—can purify their participants. Their aim is to expel evil forces at work in the world, and to revivify the salubrious and the good.

The phenomena from which a society must be purified vary hugely, and in religious contexts these would be somehow associated with the supernatural. In the secular arena of OpRealLove the target is the commodity fetish. And yet the specific form that it takes demonstrates the onion-skin quality of political life, as structures of power and their immediate opposition, though distinct, frequently take a similar shape. Peterborough's long twentieth century is a case study in how capitalism in Britain has changed. From being a place where bricks and sugar and engines were made, it transitioned to a service economy, and is now increasingly dominated by the mechanisms of distribution. Anons miniaturize some of these large-scale shifts in political economy. OpRealLove is itself a ritual, not of making, but of *distributing*, although of a very different kind to that undertaken by Amazon et al. As will be more fully explored in chapter 8, through such events Anons rehearse and reinforce their identity as a We, but taking place in the epicenter of Britain's cities, they are also performances staged for the benefit of an audience. In the following chapter we turn to face this audience. Who exactly are the You, and how do Anons engage with them?

GROWTH

I'm going to give you some love because I reckon it's going to spread.

Anon

There is a turning point in V for Vendetta when the population starts becoming mobilized. Evey stumbles on a pre-teenage girl in glasses, spray-painting the "V" symbol onto public propaganda. Boxes upon boxes of Guy Fawkes masks start mysteriously arriving at people's doors. The government is panicking. These portentous, jagged scenes, are spliced with another slower and more placid narrative unfolding. V's leather-gloved hand is carefully assembling line after line of vertical dominoes. The police character narrates over the images, "It was like a perfect pattern laid out in front of me," he says, "and I realized that we were all part of it."[1] When the girl is shot down in broad daylight, it provides the spark for riots to erupt. V's hand descends on the outer edge of one of the lines, animating the whole cascade with a simple flick of the finger. Gradually, inexorably, twenty thousand dominoes descend into the shape of a vast, encircled, V.

As with traditional initiations, digital initiation rites transport initiands from one nexus of social and material relations to another. Through the secret knowledge they have acquired online, Anons become increasingly aware of a yawning chasm that has opened up between themselves and a variety of others, an awareness which normally prompts the desire to close it. In this chapter I explore the history of the ideas that underpin these endeavors to "evolve" society, and some of their social effects.

The modality of awakening can be broadly split into two camps—a split sometimes manifest in the same person's ambivalence. In one camp, it comes in the wake of an eventful and violent conflict, thought of as revolution. In the other, change happens much more slowly, and without any violence at all, through the edifying spread of knowledge sometimes called *e*volution. These alternate visions are presupposed by precedents in Christian eschatology, known as pre- and postmillennialism. Both doctrines share a concept of the Millennium, which is to say the arrival of the end of the world, but in each it arrives in a very different way. Premillennialism conceives of end-times in terms of "high drama," as a great bloody battle between good and evil that will precede Christ's return.[2] Postmillennialism, by contrast, conceives that this return will be ushered in through the increasing perfection of the world. This second doctrine, of human perfectibility on a massive scale, came to infuse the New Age movement that emerged in the second half of the twentieth century, and it is largely through New Ageism, rather than Christianity per se, that it percolates into AnonUK. This second camp can be considered the majority view, and as the only one with tangible results, it is toward it that my focus is directed.

In the process, Anons report schisms with partners or spouses, with parents, friends, and even children and/or grandchildren—all of whom may now be subject to forms of instruction. Their engagements with other members of the British public sphere can also be transformed. In both their physical and virtual fora, Anons engage in educating efforts to awaken strangers, in which this community of non-intimates is often coded in national terms. Developing these themes, I pursue an extensive characterization of a woman named Betty, one of the most committed people to populate AnonUK. Betty describes the arrival of Anonymous as a calling, as something she had been waiting for all her life, internalizing its political ideas and practicing them on a daily basis. As an illustration of where these profound forms of subjective change may lead, I return to Betty's story again in the conclusion.

Many cults have been observed to have deleterious effects on existing intimate ties.[3] Indeed in numerous cases, cults explicitly take on the language of kinship and encourage members to substitute existing kin relations with the fictive kin of other members. What is less widely known is that it was the historical victims of these various forms of alienation—whether parents, partners, or friends—who are partly responsible for the negative connotations that presently surround the word cult, as the so-called Anti-Cult Movement (or ACM) that emerged in the 1970s and 80s was propelled by the stricken intimates of those who had flocked to the new religious forms. In its capacity to produce relationship schisms (and indeed its discourse of fictive kinship), AnonUK is continuous with some of these antecedents. Not commanding the same organizational infrastructures, however, it is possible that in the main these schisms are less categorical and have comparatively less far-reaching material consequences.

Metaphors of Nature

Anons seek to transform the way that others think about the world. Drawing partly on the filmic sequence above, it is sometimes called "the domino effect" and is explained as follows:

> You've got to convert one person. This person will convert one person, and it will just keep carrying on. You tell him. You tell her. Even him. Even kids, they need to know.

The domino effect is a metaphor for both individual and social change, each possessing a different character and velocity. Each individual domino is toppled by the one before it, undergoing the perpendicular transformation referred to as "waking up." As an aggregate however, the dominoes descend at a more even, gradual pace, eventually falling into a coherent collective shape. While Anons present their own narratives of waking up in eventful terms, in this postmillenialist vision, change happens slowly and requires a great deal more patience. For instance, when a person grumbles at the small turnout for a protest, when only a fraction of those who announce they are "going" on Facebook attend, another adhering more firmly to this ideal will display a remarkable equanimity. The argument will run that the important thing is that *they* are there, as this creates the possibility that passersby will subsequently search for Anonymous, or the event online, and through this contribute to this gradual transformation.

Anons use these natural metaphors to denote a kind of inexorability to political change. In its concrete form, the domino effect is an outcome of the laws of physics. It arises through Newton's first law of motion, when an object moves after being subject to a force, in this instance producing a chain reaction. Other natural metaphors used by Anons draw on biological growth dynamics. Another common conjured image is the scattering of seeds. One mother describes this as follows:

> I've got four children ranging from twenty to thirty. My thirty-year-old is now just starting to become aware and waking up to it, but the three younger ones are still, "Oh mum" and the eyes glaze over when I start talking. But something I have learned, people have to find their own way . . . You plant a seed and they might dismiss it initially, but then something will happen and they'll go, "Oh that's what Mum was talking about." So that's all I do, I plant a seed. It's up to them whether they decide to nourish it and bring it to fruition.

In contrast to the chain reaction of a line of dominoes, the planting of seeds possesses a more unpredictable quality—not all seeds grow up into plants—and is
therefore more dependent on the interlocutor's volition. As another says,

> I was once completely unaware of all this stuff and, if I can give some-
> body that little seed of information that makes them go, "Oh actually
> I'll have a look at that," then that's what I will do rather than preach,
> preach, preach.

Here the transmuted form of evangelism implied in these rhetorical efforts
becomes more obvious, as well as the self-conscious distinction from conversion
religion. By delegating the acquisition of knowledge to digital technologies (the
seed in a pragmatic sense being the keyword(s) entered into search engines),
Anons avoid, at least nominally, the promotion of a doctrine.

The third and most pervasive natural metaphor which frames their visions of
change is the concept of awakening from sleep that is ubiquitous across mamma-
lian life. The individual narrative of arrival into consciousness, unfurled across
this book, is understood to affect society at a large scale, through the achievement
of a critical mass of similarly conscious beings. This is the "evolution of consciousness." It may also be called the "great awakening," the "global spiritual enlightenment," or most often, a new shared "awareness." Journalists and commentators are
often baffled by what they consider the lack of purpose of AnonUK protests, but
this impression stems from a misattribution of what change for Anons consists
of—namely, the project of awakening. In answer to the question of why they have
come to the Million Mask March, one activist clarifies this project explicitly:

> To help raise public awareness about what governments are doing to
> the people around the world . . . That's why I'm here today. It may not
> change anything being here, but it certainly will raise the public aware-
> ness of what our governments are doing to their own citizens.

The total absence of telos is vivid. While effectively capitulating on the question
of any direct political consequence arising from the event, the respondent maintains the value of raising awareness as a viable long-term strategy.

Moral Evolution

The notion that a moral society will naturally emerge through epistemic shifts
on a grand scale has a deep history in European thought. The earliest known

expression of this theory of human "perfectibility" is widely attributed to Pelagius, an Irish monk writing in the fourth and fifth centuries.[4] Pelagius presented what were at the time two radical theological proposals. The first (which contravened the doctrine of predestination) argued that human beings could become virtuous not simply through the grace of God, but through their own efforts. Meanwhile the second (which contravened original sin) suggested that human nature was inherently inclined toward goodness, making a moral society the natural outcome of these endeavors. Pelagian thought thus contains the contradiction that remains present in the thought of Anons—that morality on an individual scale is a choice, while on a societal scale an inevitability. Critically, Pelagianism paved the way for Enlightenment theories of progress. Writing during the French Revolution, Antoine-Nicolas de Condorcet secularized the perfectibility thesis, linking it explicitly to the rational accumulation of knowledge.[5] Identifying ten stages of human enlightenment, from primitive forms of social organization to modern scientific discovery and invention, he maintained that civilization would incrementally, and irreversibly, advance, through increasing scientific understanding. Condorcet was an early advocate of universal education, which he believed would hasten this advance by expanding access to scientific knowledge.

The most direct source for the entry of these ideas into the political visions of Anonymous, however, comes not via Christocentric political philosophy, but via New Ageism. In its first expression, the New Age movement was a loosely defined spiritualist revival that burgeoned across Europe and North America in the 1970s. Its adherents broadly maintained that the world was witnessing the closing of a dark, violent age, known as the Piscean, and the dawning of a new Age of Aquarius, an era of love and light ushered in by a global spiritual renewal. New Age ideas have experienced a significant resurgence in the digital era, and their traces are visible across AnonUK's in-person and virtual cultural life. New Ageism is manifest in the ubiquitous discourse of love and consciousness, and in linking to websites that draw explicitly on New Age philosophy.[6] It is also manifest in the Anonymous flag (see **fig. 9**). The flag's symbolic nod to the United Nations echoes one of New Ageism's early precepts, as the UN was the preeminent institutional structure that would bring about this spiritual dawn.[7] It is also manifest at one point anthropologically. During an online radio program promoting one of their events, two prominent British Anons appear alongside leading New Age writer Barbara Marx Hubbard.[8]

It is subsequently worth diving more deeply into the origins of New Age philosophy, to establish exactly why it is so compatible with Anonymous. For this we may turn to one of its most influential intellectual sources—Teilhard de Chardin.[9] De Chardin (1881–1955) was a French Jesuit priest and a paleontologist/geologist who sought to fuse his underlying Catholicism with the discoveries of modern science, particularly Darwinian evolution. Adapting the biological concept of the "biosphere," the layer of living creatures on earth, de Chardin proposed the existence of an accompanying "noosphere," a "mind-layer" that was equally planetary in reach.[10] For de Chardin, every particle of matter possessed a psyche, however rudimentary, and with the advent of higher mammals and homo sapiens, the increasing cerebralization of matter had led to an increase in the complexity of the noosphere. In contrast to Darwin's evolutionary model which produced biological divergence, this evolution of the noosphere led to ever greater convergence, driven by an irresistible pull that he names "radial energy," and later simply "love."[11] This is where the discourse of consciousness comes in, as the final stage of evolution would come about through the noosphere's reflection on itself. Humanity and all living creatures would ultimately become aware that they constitute one entity, a "superhuman organism" or "planetary consciousness" that would then achieve a quantum moral leap by behaving as such.[12]

The facility with which these notions of planetary consciousness can be fused with contemporary methods of virtual communication is not entirely by chance. De Chardin considered the global expansion of telecommunication systems in the 1940s and 50s as integral to the development of the noosphere.[13] There is no way he could have foreseen, though, the sheer ubiquity and simultaneity of digital communication in places like 2010s Britain, and in this sense there is a remarkable coincidence of fit between his vision of a mind-layer which operates out of, but to some extent autonomous from, biological life, and the actual technical modalities to achieve this. When Anons update this foundation of New Age thought to the affordances of twenty-first-century technology, their explicit emphasis is on the role of hegemonic social media in expediting this spiritual shift. Pursuing the vein of the perfectibility thesis, Anons suggest that these platforms will play a critical role in "connecting people," and "actually help our consciousness progress." For instance, **figure 11** shows a homemade graphic in which the Facebook invite function is imagined as a way to achieve the domino effect.

Before examining how these ideas inflect a range of social relationships, we can observe two major lacuna in this conceptual system that hinder its execution. The first is the absence of recognition that computer-mediated sociality is not a plant or a play of forces, but a highly structured terrain cross-cut by a variety of corporate and state interests.[14] The role of personalization and its algorithms in determining what Internet users see, and who they are connected with, has become more widely understood since AnonUK flourished, and in their undiluted cyberutopianism, Anons display their backward glance to the years before this economy of Internet surveillance existed. Second, Anons maintain that this mass awakening will arise through the spread of information to which the You

FIGURE 11.   Imagining the Facebook invite function as a form of political action

will rationally respond. Yet in their own narratives, it is persistently their subjective knowledge of invisibility that lays the groundwork for awakening to occur. Although there may be an inchoate sense that, as the mother of four says, "something will happen" that allows the seed to germinate, there is no theory, nor process, through which such happenings are brought about. This is one of the key differences between traditional and digital initiation. While the latter's efficacy arises through the same subjective sequences of death and rebirth, these are not ritually orchestrated and are therefore essentially unpredictable.

Changing Relationships

"Initiation defines boundaries," Jean La Fontaine says.[15] In ritual contexts it ministers transitions in the social status of the initiand—from child to adult, from novice to master—rewriting the social and material boundaries which shape their place in the world. It is for this reason, she argues, that these rites are often concluded with large public ceremonies, as the change they bring about affects not only the initiand, but every other person with whom they are connected. In this digital initiation rite, the change is publicized through a voluntary identification with linguistic terms and symbols rather than a large ceremonial (although as we see in the next chapter, sacrificial rituals consolidate the shift). Yet the boundaries reordered by initiation remain profound, perhaps even more profound because they have *not* been ceremonially performed. In the course of waking up, Anons discover that all of their relationships have somehow changed, from the closest intimates to the most distant strangers. This change is met with a range of responses that are specific to the relationship in question.

Intimates

Waking up can instigate a schism with a wife, or husband, or romantic partner. In a quantitative sense, time now engaged in political activities (exchanging on social media or going to events) may be time formerly spent with this person. Of larger significance is the schismatic reduction in quality, as the knowledge acquired on waking up alters long-term habits to which the relationship must adjust. These frequently manifest themselves with reference to consumption. For instance, a new insight into the meat industry may entail their becoming vegetarian, forcing new commensality arrangements. Media consumption can be particularly sensitive. One Anon, upon learning that television personality Simon Cowell had lent financial support to the Israeli Defense Force, decided to boycott his flagship program—X Factor.[16] It was something he and his wife once watched together and even looked forward to each year, but while its meaning had radi-cally altered for him, this was not the case for her; she still took pleasure in it. As a compromise she continued to watch the program downstairs, while he went to their room upstairs, to watch consciousness-raising documentaries instead (and establishing a Facebook group around the project). In most cases the relationship can withstand the change, the partner in question referred to with affectionate resignation as a "total sheep." It is possible, though, that waking up can yield a transformation so great and so enduring, that it leads to separation and divorce.

As with many of those who joined the traditional cults, waking up can deepen existing schisms with one's parents, particularly fathers and stepfathers. This is more often the case for British Anons. While those born and raised outside the country are more likely to cite leftist parents whose way of thinking still informs them, those from Britain generally describe their parents in right-of-center terms—such as reading right-wing newspapers and voting Conservative. The left step that an association with Anonymous can produce, through its particular repudiation of possessive individualism, can upend desires their parents took for granted. As in relations with romantic partners, this widening gulf may prompt the desire to close it and recover an intimacy lost. These are seed-scattering efforts, though, rather than doctrinal diatribes, and rely substantively on digital technologies. One Anon described an attempt to wake up his father:

> He doesn't agree with anything I do. He won't even consider the fact that we live in a society that is unjust and corrupt. He just doesn't want to see it. And I tried to show him. I tried to show him the Matrix scene with the blue and the red pill to say, "What this is actually about is people waking up to what is really going on in society, not just what we're told." "I don't want to see that" (he says). He just doesn't want to see it. He wants to carry on living in the system, this safe kind of world. I mean all right, he's in his 70s now, but the point is he's never wanted to wake up to what is really going on and he's never questioned it.

Adopting his perspective, there is a pathos of separation, and an earnest long-ing to bring his father around to his altered way of seeing. Imagining his father's viewpoint, however, one can see the obstacles. A retired septuagenarian is shown a scene from a science fiction film as evidence that the world is not what he thought. The encounter for him may have been equally exasperating.[17]

The distancing of intimates is a key stage of initiatory passage. In ritual contexts distancing is actively produced and controlled, taking place through conspicuous performances of separation. Because so much of this literature addresses male puberty rituals, the most well-documented kind is the separation of the nov-ice boy from his mother. These can be brutal affairs. Masked men arrive in the middle of the night and rip the boy from his mother's arms, carrying him into the forest to the sounds of the latter's wails and pleading.[18] They can be rather calmer, mother and son acting out the new distance between them through bodily gesture, or the removal or display of ornament.[19] It is impossible to say what the mothers are feeling at these moments, but there is often a performance at least of high emotion, during which they weep loudly, or smear themselves in ash to express their grief.[20] Grief is not necessarily too strong a word as for all intents and purposes the boy has died to childhood, and their relationship may be very different from that point onward. This usually entails new living arrangements, as the boy is removed from his mother's care to enter the secretive world of adult men.[21] In the more extreme cases, he may be compelled thereafter to consider his mother no more meaningful than any other woman, or even forbidden from talking to her ever again.[22]

These digital initiations involve a very different attitude toward minors. Indeed, it is Anons' subsequent relationships with their children and grandchildren that bear the least evidence of separation. From existing accounts, it appears that the first individuations of Anonymous possessed an overwhelmingly priapic character, populated by young adult males in their twenties and teens. Alongside the arrival of women in greater numbers, AnonUK also witnessed the arrival of children, and many of the events described number children under the age of ten among those present. This is particularly so for the rituals of OpSafeWinter, sometimes advertised as "family friendly" occasions, and which deliberately include activities for kids. Even if they do not, though, children take immense delight in the ludic spaces of Anonymous, with its oral exuberance and its material culture of masks and props. The inclusion of children in AnonUK is more than just a solution to childcare. It serves a demonstrative function, inculcating these young people in its values. Like the girl character in *V for Vendetta* who is placed at the center of its revolutionary transition, children are considered a vital part of the "domino effect"—perhaps even the most vital given their greater epistemic plasticity.

Strangers

Building on its Christian precedents, New Ageism popularized the application of intimate terms and behaviors to non-intimates, and this move endures inside Anonymous. Acts of "love" and "kindness" are understood to ameliorate society's moral core. This has two practical outcomes for their interactions with strangers in the public sphere: the "Random Act of Kindness" and the "Free Hug." The Random Act of Kindness is an act of impromptu generosity meted out to the unsuspecting stranger. It is also a hashtag and can be used to watch dozens of these deeds online. Those of the most interest to Anons turn on the more spe-cific relation between the urban pedestrian and the static urban homeless, the latter generally visited with gifts of food or money. A Free Hug, meanwhile, is the full-armed embrace of a stranger (though they may also be exchanged between friends).[23] That it is free implies it is a gift, but also suggests an element of trans-gression, as its invitation to physical contact expresses open disregard for exist-ing behavioral norms. Indeed, it seems possible that the Free Hug is a lexical twist on the "Free Love" of early New Ageism, which was also transgressive in its valorization of sexual promiscuity. Although the Random Act of Kindness and the Free Hug occur across other global subcultures, their particular expression in AnonUK speaks to its particular theory of social change. Both are character-ized by a fleeting spontaneity that substitutes for any form of telos, and both are efforts directed toward individuals, in particular fellow urban subjects. I enjoy many Free Hugs over the course of fieldwork, and at their tightest moments, they do transmit the sense that loving a complete stranger is within the scope of human possibility.

These intense displays of intimacy between strangers have been called love-bombing. Eileen Barker, in her description of the recruitment workshops run by the Unification Church in the 1970s and 80s, describes scenes in which members overwhelm their guests with affection, or stop everything to cook them dinner.[24] Barker herself records being presented with a rose for being the "beautiful person of the evening," only later realizing that these roses had probably been presented to many others.[25] Barker addresses the question of whether these simulations of love are simply recruitment tactics. In the workshops, they took on an immersive quality, as they were combined with the inability of guests to speak with one another alone without a member present—which certainly suggests a degree of strategy. But Barker also holds open the possibility that for some, it was through love-bombing that the Kingdom of Heaven on earth that they sought would be restored. In AnonUK, these different motivations cannot be satisfactorily detan-gled either. Both Free Hugs and Random Acts of Kindness, for them, exhibit moral evolution, and also, particularly when they are filmed and disseminated online, promote the joining of Anonymous.

Another way of attempting to engage with strangers is through humor. As outlined above, in prior forms of Anonymous this was codified as the Lulz, a form of humor that thrived on the misfortune of others, deliberately stepping on cultural taboos.[26] Here this cruelty largely disappears (in the physical public sphere at least) leaving its more benevolent, prankish side. Like the other acts above, it also contains an element of transgression that is justified by the need to puncture the veil of slumber that surrounds this demographic. One female Anon recalls with laughter an episode during which her fellow Anon reenacted the legend of Lady Godiva. Godiva was an eleventh-century English noblewoman who allegedly rode naked through the streets of Coventry covered only by her hair, to protest her husband's tax on the peasantry.[27] Equally nude except for a blonde wig, the woman in question staged the performance to raise awareness of contemporary fiscal inequalities. As she recalls,

> We ended up with a crowd of sixty people listening to exactly what she had to say, people who would not have stopped to listen otherwise. All right, it was a stick horse, it was a wig, she did look more like Cousin Itt—but it worked.[28] Those sixty people learned something they didn't know before, before we were chased out of Coventry by the police. It's little moves like that, making it outrageous enough so that the public on the outside can hear.

The idea that the public could not "hear" prior to that, extends the metaphor of sleep beyond the ocular. Moreover (much like the way the Million Mask March is inscribed onto the deep ritual history of bonfire night), the decision to draw on an Indigenous folkloric culture to make this satirical intervention codes both actor and audience as English.

It is not all love and laughter. The sense of frustration and defeat that can infuse efforts to wake up their intimates may equally saturate relations with strangers. Consciousness-raising exhausts a large amount of energy, and there are moments when reserves are running low. One Anon who had spent time in the British Army compares this feeling to having "twenty million blanks fired at you." You know it is not going to hurt physically, he explains, but it certainly hurts psychologically to be denied in such a way. This metaphor communicates another aspect of their organicist vision. Because Anons are not evangelists *strictu sensu*, they can be pacifist in the face of contradiction. Steven Sampson describes the ways in which evangelizing efforts by members of a cult can suddenly switch from warmth to aggression, when it becomes apparent that their message is falling flat.[29] Anons' relations with strangers rarely exhibit this aggression (if anything it is the other way around) but are instead more likely to descend into an enervated patience. Because the You are always potential recruits to the We—being people who simply do not "know better yet"—they do not inspire the anger reserved for the inimical They.

Whether with strangers or intimates, the success of these endeavors is uncertain. To extend the seed metaphor, this is not the labor-intensive work of mass farming, sowing seeds on a large scale in soil that has been prepared to ensure a good volume will flourish. It is work with minimal labor costs, and far more spontaneous effects. In the years 2014–15 Anons tend toward an exaggerated optimism. They are sure that those standing on the sidelines of a protest are going to subsequently "look it up" online, or that there is complete symmetry between their message and its reception. The Anon above asserts that those gathered around Lady Godiva were listening to "exactly" what her friend had to say, but without concrete evidence. It seems plausible that at least some of them were dazzled by the sight of a naked woman in an unfamiliar setting. With adult intimates, Anons have a stronger sense of their inner lives, and their energies in this direction are more likely to be judged as failures. Those most potentially likely to be imprinted by their theories are their children and grandchildren, given the asymmetrical quality of these relationships, but it will take a generation or even two to understand what plants these seeds grow into. Like the seeds Anons sometimes scatter in their own gardens, some may remain dormant while others shoot up into big bushy shrubs.

Friends

The third relation in flux is friendship. Newly identifying as Anon can produce alienation from one's former friends, to the extent that even decades-old friendships may come to an end when it becomes apparent to what extent their values have shifted. Quickly filling this void, Anons will commonly enjoy a vast proliferation of new friends through Anonymous, at least in the way that Facebook has defined the term, in other words, in a way which includes people they have never personally met. As one puts it, they lose "friends," and gain "acquaintances." Unless they are resident in one of the major cities where more Anons cluster, many of these new friendships will be structured by what sociologist Mark Granovetter calls "weak ties."[30] A weak tie is defined as a relationship between person A and person B that is not triangulated by any person C. It is when someone knows someone, but does not personally know any member of their network, and they do not know any member of theirs. In his typology it is counterposed to the "strong tie"—when these networks overlap significantly.[31] Granovetter's most salient conclusion is that it is weak ties that are of greater macrosocial significance, as they provide singular links across vast network changes. When Anons' efforts at moral evolution are re-mediated across digital channels, they serve to reinforce the strength of the weak ties that have produced Anonymous as a macrosocial phenomenon. Granovetter's theory goes some way toward explaining how a relatively small group of geographically dispersed people managed over this period to achieve such a high degree of public prominence. With this in mind, I conclude with a portrait of one of Anonymous's fiercest advocates, Betty.

Betty

A few dozen people are gathered across the pavement and road outside Parliament. They have made their way here from Trafalgar Square as part of an ambiguously titled "March for Freedom UK," and the speeches unite in a generalized critique of the British state. At the edges the gathering folds outward. Whenever a double-decker bus looms past from the nearby traffic lane, several of those at the back turn and wave at its occupants, shivering their placards and smiling. Of these, Betty is the most assiduous, swiveling often. She is also the most glamorous: platinum blonde hair tied back beneath a white fedora, all set off with a shiny black puffer coat and scarlet lipstick. The Guy Fawkes mask on top of her hat has also been visibly feminized, eye-holes embellished with thick black lashes, and lips painted red in emulation of her own. When one of the speakers enjoins those present to exchange contact details, Betty instantly extends her right hand in my direction, into which I drop my university card. The casualness of that gesture draws me toward her. Displaying none of the hesitancy around strangers that can characterize metropolitan exchange, Betty says simply on inspecting it, "We invited the students union," and starts talking about the expanding role of private security firms in the British police force.

Even at the age of sixty-one, Betty dissolves easily into different kinds of protests. At a demonstration against the blockade in Gaza she arrives with a keffiyeh wrapped around her head, her arms splashed in streams of red paint to represent bloodshed.[32] On another day she visits a woodland camp in Greater London—where activists are squatting to obstruct a property development—and drops off four heavy shopping bags loaded with processed foods and cooking utensils. One of the most consistent objects of her political energies is her 86-year-old mother, who lives on a council estate in London's West Hendon. Her mother bought the flat from the council in the 1980s, although they still manage it, and Betty has come to her aid over a number of maintenance concerns. Most recently the estate has been targeted for "regeneration," which in practice means the council is requesting the property back so that the estate can be demolished, while offering her another at a much higher price.[33] Indeed when a journalist interviews her mother for a news feature about the scheme, Betty sits there beside her, quickly filling in the gaps when her mother's words falter. The symbolic and practical portability of Anonymous's iconography suits Betty perfectly, offering a wider context for her individual style of activism. She says she takes her mask and flag with her everywhere she goes.

Betty is able to dissolve easily because she herself is insoluble. Her identification with AnonUK has amplified an already robust sense of self and given her a vision for change that she seeks to bring others into. This might take place during informal conversations—someone coming to fix her tap will get a "dose" of what she knows—while on other occasions it takes place through gestures, when she consciously performs acts of generosity to strengthen the domino effect. Those in which she manifests most delight, are the ones that transgress a categorical boundary. She is particularly fond of "educating" and embracing members of the police force, uploading one image thereof as her social media profile. She also describes with great tenderness a memorable encounter at the OpSafeWinter event she administers. A young woman wearing a headscarf, whom she under-stood to be Muslim, approached the table gingerly, taking some of the items for donation. In return, Betty told her, she would like a hug. The woman initially hesitates, stepping back, "And then the biggest smile ever came across her face like, oh my God, this White woman wants to hug me, a Muslim." Such emotion-ally charged episodes are the most generous expressions of her politics.

Betty lives in a terraced two-bedroom house just off a highway in Greater London. Outside you can hear the metallic thrum of passing vehicles, but inside it is neat and quiet, and she asks me to take my shoes off while reaching for some filtered water in the fridge. It is a hot day and I receive it gratefully. Her living space functions as an open plan: a kitchen a few paces across, a circular wooden dining table, and a lounge area with two sofas which meet at right angles around a low-slung table. Diagonally to the sofas is a nook where the ceiling curves down to meet a modest desk space. On it sits a desktop computer, with newspaper cut-tings and flyers for the Million Mask March affixed to the wall just above it—all watched over by the eyeless gaze of a suspended Guy Fawkes mask. After hand-ing me the water, Betty says I can use the computer to go online if I would like, to help myself feel at home. I demur, and eventually we settle down on the sofas to watch *V for Vendetta* at her behest, through a large-screen TV. She must have seen it more than once as she interrupts our chatter over the film to exclaim "I love this bit" at key moments. While the light outside fades to black, Betty nimbly rolls herself a large spliff which she then proceeds to smoke. As she refuses any pharmaceutical drugs, cannabis is her primary form of pain relief.

Betty contracted polio at the age of three during the 1956 epidemic. It is a virus which attacks the nerves in the spine and the base of the brain, and can leave the body with weakened muscles and aching joints. Her post-polio health is irregular. Some days she can walk without the aid of a stick, and although at protests she is never without her exuberant pink wheelchair, she is also able to stand up if she wants to. On other days Betty is forced to spend hours lying on her back, whiling away the time watching YouTube videos and following protests on her smartphone, and at these moments her spliffs are a great source of comfort (like many Anons Betty is evangelical about the healing properties of cannabis). Her own story of waking up is closely linked to her own troubled health. Besides a switch from designer clothing and junk foods to secondhand wares and some home-grown vegetables, her most dramatic transition has been to stop taking the panoply of pharmaceuticals she was prescribed for her condition. Now, she says with pride, besides cannabis she will use other natural remedies such as onions, turmeric, and garlic to cure her ailments, rather than the "poison" provided by the medical profession. In Betty's narrative, the failure of care that produced her separation came at school. She attended Catholic schools in the 1950s and 60s and describes the elders subjecting her and her classmates to corporal punishment and "psychological torture." This experience provides much of the fodder for her own impassioned denunciation of institutional religion, particularly Catholicism.

Yet it is also the case—as she openly acknowledges—that these years of religious schooling exerted an enormous impact on her, as her speech is shot through with theological notions. The first formal conversation we share takes place in a café in North London, one of those easy-going establishments with yellow lighting and wipe-down chairs. Besides "another Anon girl" who joins us in the discussion, Betty brings along her teenage grandson who, like her, is dressed entirely in white. He is smiley and self-contained, fidgeting on his smartphone and whatever else is around as we talk. "I've always fought for people," she explains,

> And then one day I came upon Anonymous and I thought, "This is my calling . . . This is what I've been waiting for all my life". . . We're not a religion, we're not anything; we're just a collective of like-minded people with one idea: to make the world a better place, a better place for humanity, a fair world.

Betty is among a handful of participants for whom the arrival of AnonUK offered a powerful vocation, giving their lives a radically new coherence where before there were inchoate ideas and disparate goals, a vocation bound up with beliefs in human perfectibility. There is one specific instance, though, which illustrates her more explicit connection to some of the New Age ideas above. Spotting a man begging in a doorway across the road, she briefly halts the flow of words to instruct her grandson to ask the man if he would like some coffee and something to eat. Obligingly he does, and returns with the order.

> Hot chocolate? What, nothing to eat? Awww poor homeless man over there. But you know that's my act of kindness for today. I try to do it every time I go out.

Her subsequent elaboration of these random acts of kindness gives the episode a performative feel, staged for the ethnographer perhaps, but this does not necessarily preclude a simultaneous didactic intention. She is also demonstrating this theory to her grandson, afterward explaining that they are both "Aquarians" who like to help people. The hours the young man spends with Betty on this and other occasions suggests his sympathy for his grandmother, and the ever-so-slightly wry eye with which he carries out her request suggests there is an element of indulgence to the interaction that is undergirded by it.

If, in the final analysis, the endeavors to awaken intimates, friends, and strangers fall largely flat—why is so much political energy expended in this direction? Here it is worth returning to Jean La Fontaine's assessment of the fundamental purpose of these rites—namely to reproduce existing forms of social and cultural authority.[34] La Fontaine presents an important counterpoint to the intensive focus on the initiand in the American psychoanalytic tradition, arguing instead that these rites bear most significance for the already initiated, because they reinforce the social and cultural value of a ritual these persons have gone through and now command.[35] In other words, the urgent need that Anons display to awaken others is driven not simply, nor perhaps even mainly, by a desire to augment the social good. The more people that rise from their slumber in the way that Anons understand it, the more this legitimates their own experiences of transformation. It is the consequently the "I" that holds the three other pronouns in place. With all forms of segmentation now described, I contemplate the rites of public sacrifice that make these weak ties stronger.

SACRIFICE

**I learned to give not because I have much,
but because I know exactly how it feels to have nothing.**

#OpSafeWinter meme

A video shows a group of Anons parading down a narrow shopping street in Nottingham.[1] Sonorous chants fill the audio, as one of the masked men at the front notices something out of frame and points to it. The camera pans down to disclose a man past middle-age in a flat cap seated at the side of the street, begging. The vanguard bring the others to a halt.

> Stop. Stop! Everybody. If you walk past this guy, and you don't give him no (sic) money, then none of this means nothing to you!
>
> This man is Anonymous!

Coins and notes flow into the beggar's collection hat, and Anons take their turn to shake his hand, some filming themselves as they do.[2] The beggar himself, who may not speak much English, keeps nodding plaintively, shoulders hunched and head bowed, adapting with expressive skill to what must be a startling flurry of activity around him.[3] Soon the men who had initiated the encounter squat down on their ankles on either side of him, and put their arms around his shoulders. More photos are taken.

> You're a good man. You know that? We love you man.

This moment of spontaneous physical and discursive intimacy seems to have a transformative effect on the beggar. His face begins to crease. Then a woman wearing a golden mask, carrying a sign advertising free hugs embraces him too, and the creasing turns to weeping. He rubs one of his eyes.

He's crying. He's crying!

The beggar must have muttered a word of thanks, because he is gently admonished for doing so. We are all being robbed, he is told. After a few minutes the happening draws to an end amidst more handshaking and patting, and the beggar is bid farewell with words of hope.

Better days are coming. Better days are coming.

This YouTube video, filmed and published in early 2013, both signaled and produced a landmark shift in what it meant to be Anonymous in Britain. In the months and years that follow, it is widely viewed inside AnonUK, shared and reshared, with some reporting spontaneously how it had moved them. The title of the film, "Anonymous makes homeless man cry," anchors its overarching meaning around this display of affect, because it materializes an experience of suffering that Anons consider continuous with their own. This shift is the point at which previous individuations of Anonymous begin to coalesce into AnonUK. From 2013 onward, Anonymous in Britain exhibits a new orientation around the urban homeless. Assembling under the hashtag #OpSafeWinter, it entails pouring a variety of resources into the care of such persons, who become the focus for the amelioration of society that is sought on waking up. This chapter chronicles one OpSafeWinter that takes place in Nottingham in late 2014, followed by a description of an eviction block that happens in the same city six weeks later. Like the previous portrayals of Manchester and Peterborough, this ethnography is preceded by a history of the city, which has for centuries has been bound to the myth of the moral outlaw.

OpSafeWinter concludes the sequence of their digital initiation rite. Taking both individual and collective expressions, it is the moment when Anons finally renounce the transitional period of symbolic death, and engage in public activities they hold to be meaningful and good. Albeit taking place in a secular arena, the character of these practices is prefigured by the long history of religious sacrifice, in which something is given up in order that something else (in this case a person) can regenerate and persist. This sacrificial subtext binds Anonymous to the moral logics of the postwar British state, and the value placed upon personal sacrifice for the sake of collective good—one that becomes weaponized in the age of austerity. In the chronology of their initiations, the provision of resources for the homeless becomes a sanctifying experience for Anons, allowing them, for a time, to expel the sense of being invisible that instigates the rite.

Nottingham

Nottingham is a medium-sized city in an area of England called the Midlands. Midlands is a particularly apt designation in this instance, because Nottingham sits like a bullseye right in the center of England. The contemporary history of Nottingham began in 1067, when William the Conqueror built a fortress on a raised area of rock above where the city now lies, which became the basis for Nottingham Castle, a significant seat for nobility and royalty throughout the Middle Ages.[4] He also claimed substantial areas of forest around Nottingham as crown land, which were then governed by a particularly punitive system of fines, taxes, and levies, whose infringement prompted macabre forms of mutilation or hanging.[5] Yet it was from deep within the same forest that the other side of Nottingham grew. In the context of a brutal feudal regime, itinerant balladeers and minstrels sang songs about a figure known as Robin Hood—a free but unproperties man—who defied the punitive regulations of the forest with his band of fellow outlaws.[6] With the development of the printing press, this oral culture of unknown provenance found its way into books, and the myth evolved and proliferated for centuries afterward.[7] Now Robin Hood is unquestionably more famous than the castle that was his nemesis, and in the 1950s, a seven foot bronze statue of him was erected beneath it, his bow taut and arrow poised, taking aim at its occupants in perpetuity.

The presence of a power that comes from the ground infuses the modern history of Nottingham, too. In the seventeenth and eighteenth centuries the city nurtured a thriving cottage industry.[8] Families wove silk, wool, and hosiery on small frames inside their homes. With the arrival of the industrial revolution, this pool of manual skill quickly allowed Nottingham to establish itself as a powerhouse of textile production, and just as Peterborough's history was molded by clay, Nottingham's was woven by lace, with its products exported to the United States, Latin America, and elsewhere. This power from the ground did not easily cede itself to the new class of factory owners. Between 1811 and 1816, it was Nottingham that incubated the practice of machine-breaking known as Luddism, as workers destroyed more than a thousand weaving frames in response to their employers' efforts to drive down wages.[9] A "spirit of riot," as it was called, was somehow nested inside the city.[10] In 1831, in rage at the Duke of Newcastle's opposition to parliamentary reform, some twenty thousand protesters marched to his seat of Nottingham Castle, ransacking it and setting it alight, its charred remains standing untouched for several decades. In the years that followed, Nottingham became a major center for the Chartist movement, electing the only ever Chartist MP. This riotous history should not, however, be straightforwardly conflated with Leftism. It has been observed that despite the disturbances around the turn of the nineteenth century, the trade union movement did not make significant inroads into Nottingham until the craft societies unionized in the 1850s.[11]

The crucible for Nottingham's political activity was, and remains, its Old Market Square.[12] It is the largest surviving market square in Britain, on which some version thereof has stood for more than a thousand years, binding the pre- and the postindustrial in place. The square consists of an expansive pedestrian area in the middle, with a line of shops along the northern side, and a tramline that curls around the southern and western sides. The square is dominated on the remaining side by the majestic architecture of Nottingham Council House. Erected in the 1920s in a grey English limestone, it has a four-story neoclassical façade capped by an Art Deco relief illustrating the idealized activities of the council, itself superseded by a two-hundred-foot-high gilded dome. The Council House can be considered in a similar light to Manchester Town Hall, which is the say that it was commissioned near the apex of the city's industrial prowess, to endow its structures of governance with an unapologetic splendor.[13] More so than Manchester, the changing utility of the building over the long twentieth century illustrates changing ideas about the role of governance. In 1985, during the first wave of privatization by the presiding Conservative Party, the ground floor was sold off and redeveloped as a luxury shopping arcade. Then in 2010, at the beginning of the austerity era, two more floors were emptied of their elected representatives, leaving only the top floor, containing the original council chamber, still used for public meetings. Outside the building sit two larger than life-size stone lions that face one another like the building's sentries. It is at these lions that Anonymous invites people to gather, via a YouTube video, for an OpSafeWinter event twelve days before Christmas.[14]

Continuing the city's historical association with outlawry, the event is organized by a group of Nottingham-based Anons calling themselves the "Pirates." The Pirates are distinct from other Anons, to the extent that they tend to be more focused on their internal collective life than on "awakening the masses." They also mark themselves out aesthetically, painting their Guy Fawkes masks half black down one side, and consistently bring their own Skull and Crossbones flag along to the Million Mask March. The Pirates are among the minority in AnonUK who still use the peculiar argot of IRC, one indicator of the greater policing of boundaries between insiders and outsiders that characterizes this group.[15] The net effect is that their sense of collectivity is more meaningful here than in other urban centers; indeed, if AnonUK did possess a heartland in Britain— Nottingham would be it. It is the de facto home of AnonUKRadio—an Internet radio station broadcasting several times a week—and it is from the YouTube accounts of the Pirates and their associates that several of the masked newsreader videos are disseminated.[16] It is also the case, in the lore of AnonUK, that it was the Pirates who first conceived of OpSafeWinter.

OpSafeWinter

OpSafeWinter is a ritual performed at any time of year, either by an individual or by a group, during which donations of money or needed items are distributed to people identified as homeless. Both the term and the concept first appeared online in November 2013. As the story goes, after that year's Million Mask March, some of the Pirates and their friends spent the night pacing the streets of London. In this activity they could not help but notice the sheer volume of people sleeping out in the cold, and they determined to do something about it. The following day they returned to the subject on IRC, and OpSafeWinter was born. In the absence of chat logs, this origin tale cannot be verified, but it is the case that the first #OpSafeWinter hashtag was published by AnonUKRadio's Twitter account on 7 November 2013, together with a link to a text on the platform Pastebin.[17] The authorship of the text is masked, but a reference in it to the "local council" suggests a British orientation, these assemblies being components of Britain's representative democracy.[18] The initial tweet is soon reposted by a number of other accounts containing additional British references.[19] It is only when the hashtag is retweeted by one of the largest global mediators for Anonymous content—YourAnonNews—that OpSafeWinter fully erupts as a transnational Internet phenomenon. On 28 November 2013, a film is published on YouTube titled "Everybody is me #OpSafeWinter," featuring emotional scenes from the Steven video.[20]

Hence when the Pirates announce their invitation to gather in Nottingham's Market Square one year later, the Op has already matured as both an idea and a praxis, charging members with anticipation. A Facebook event page is set up to provide a space for the exchange of ideas and information beforehand. Two aspects of these exchanges stand out. The first is the pervasive presence of an afore-mentioned New Ageist discourse of intimacy as applied to non-intimates. One participant proposes to inaugurate the event with an hour of Free Hugs.

> Its time we started to love each other. all around us there is hate. hate for poor, hate for rich, hate for colour, gender, beliefs. no more I say. we will love. join us and give free hugs to any of those that want them. unite under the one unifying banner. the banner of love #FreeHugs[21]

This is subsequently reinforced by references to "love flowing," of "love to all," a profusion of heart-shaped emoticons, and images of people embracing. Second, something that is definitional to OpSafeWinter when and wherever it appears, the event is fully enveloped within a language of humanity. Being human is represented as something innate to every person, but that OpSafeWinter specifically allows them to tune into. The implication is that being human is inextricably linked to care.

> We are all human beings & we were initially designed to help one another

Alongside this, interlocutors frequently refer to one another using fictive kin terms, as brothers and sisters—or indeed "bro" and "sis."

On a bitingly cold Saturday in December the hour arrives. Yet it does so in a free-form, incremental way, rather than with any great fanfare. As proposed there is much embracing, but the free hugs promised online are predominately shared between people who know each other already and are wearing signifiers of membership. Like their allies in Peterborough, the Pirates have had high-visibility vests custom-made, in this case with the words "OpSafeWinter Nottingham," which they wear with Guy Fawkes masks propped in thanatoid style. Besides the vests it is clear that the event has involved manifold preparations. Someone has sourced a dozen military grade sleeping bags. Someone else has invited a vegetarian catering company to distribute hot food. Even just that morning, the Pirates had been up early assembling the bags of dried consumables and second-hand clothing that form the backbone of any OpSafeWinter, bringing it to the square in a people carrier. Yet none of this effort is visible now. A woman wearing a Free Hugs sign over her fluorescent vest, and a Guy Fawkes mask with a pirate's patch penned around one of the eye-holes, tells me, "All the organization happened beforehand. There's just no organization on the day." The presence of work and planning is deliberately excluded from the rite, creating a space that can be filled with something else.

Instead of practical work, the hours glide by like a "moment in and out of time."[22] As homeless people come and go, foraging in the bags for winter clothes and receiving free food, Anons stand around chatting in small groups, their attention collected at intervals by fleeting social dramas. At one point, two male Anons jump astride one of the stone lions, chanting, "Help the homeless!"—a chant which is immediately echoed by those standing below. At another, a group of eight or so Anons dance on the shallow steps of the Council House to bass-heavy music coming from a small speaker system. It is a specific way of dancing that became popular in Anonymous during its anti-Scientology period, which consists of taking a wide-legged stance and slightly bending one's knees, hugging elbows to the torso with hands pointing out, and then waving and wiggling like a wooden top from side to side. As well as a moment of embodied collectivity, it is a performance intended for the benefit of future online audiences. Before the smartphones start filming, several of the Anons switch modes from revelation to concealment, pulling their masks down over their faces. What most distinguishes these masked dances from the others that take place during initiation rites is their intrinsic unseriousness. In traditional rites there is normally a solemnity to the emergence of masked beings.[23] But in Anonymous, there is a clear communication being made—this is a joke.

It should be added that just because something is represented as a joke, does not necessarily make it benign. In fact it may be precisely the opposite. Taussig notes the entangled relationship between humor and violence, and not long afterward, on the very same spot, one morphs into a version of the other.[24] In a modest effort to inscribe the space, the Pirates suspend a large green monochrome Anonymous flag between one of the building's grand arches. This prompts a man wearing the burgundy regalia of the Council House's private security contractor to surface and request that they remove it. The benevolent spirit that has framed the day so far evaporates with remarkable speed. The Pirates round on the guard, shoulders pulled back and voices raised, objecting vociferously to the request. This quickly grabs the attention of those standing by and stimulates a moment of hyper-mediation. Phone cameras are drawn like swords, and the guard and the colleague who has now joined him are warningly informed that they are "going on YouTube." In spite of all this, however, the flag is ultimately taken down, and the event continues almost as before.

OpSafeWinter materializes an ethic of self-organization that was central to the forms of cyberutopianism that generated Anonymous. It is in essence the idea that social life can propagate itself in healthy and effective ways in the absence of leadership, or any kind of command system.[25] The Pirates emphasize throughout the day that they "didn't need to do anything," which, while overlooking all the work the event does involve, speaks to their desire to incarnate this ideal. In AnonUK, the forms of lateral collaboration documented by Coleman online take place within a stronger theory of the human. Here being human entails a natural inclination to support those "weaker than yourself," which is permitted room to bloom when all hierarchical structures are dismantled. Closely tied to this idea of care emerging in the absence of work or structure of any kind is a conception of the distribution of goods among the urban homeless as strictly nontransactional. It is critical to the proponents of OpSafeWinter that the items they provide are not given, but are instead taken by those who require them in conditions of autonomy. This is illustrated when one Pirate parent points to those searching in the bags for clothes and explains to her young son, "Look, Gary, the people who need the stuff are helping themselves."

By late afternoon, the December darkness descends, and it is time to pack up. All of a sudden, the energy it took to instigate the OpSafeWinter becomes visible once again. Men lift up binbags of clothes as though they are filled with feathers, cramming them into the people carrier that has returned to the side of the square. The plan is to deposit them at a local homeless shelter for further distribution. When participants say their goodbyes on the immaculately cleared slabs, it is a moment of reflection and grace. Several remark with wonderment how the scale of the event has increased many-fold upon the previous year, while individuals known for trolling look out on the world with a sheen in their eyes. The hugs are again exchanged with one another, except this time they are longer, more meaningful. The sense of enchantment persists for many days afterward, one participant posting online,

> On my way home from one of the best weekends of my life. The real feeling of love and empowerment that was giving so freely is like nothing I have ever experienced. The bonds that have been made this weekend will never be lost.

It is an event that floods the Pirates with affect, and through this strengthens their connection. But how does such an occasion achieve its spiritual efficacy? And why is this the main practical consequence of waking up?

Sacrifice

OpSafeWinter is the act of giving something up. What is usually donated are material resources—money, food and drink, clothing, toiletries—but it would be a mistake to consider OpSafeWinter in purely material terms. In uneven ways Anons give up their time, their energy, as well as the emotional resources demanded by any authentic relationship of care. We could return to Kate's interaction with Mike chronicled in chapter 2. This involves not only a material intervention (a pair of shoes is found, an ambulance called), but perhaps more importantly, a physical and emotional intervention in the offering of a shoulder to lean on, and a labor of sympathy more generally within which the episode unfolds. In giving something up for the benefit of homeless people, Anons make conscious and deliberate efforts not to look down. When Anons, like Kate, interact with the destitute sitting on pavements, they tend not to bend from the waist keeping their legs straight, but rather to rest their torsos on their haunches with knees in a tight squat, so that their heads and spines mirror the geometry of their interlocutors. This willful leveling is also expressed by another hashtag that surrounds these events—#SolidarityNotCharity. The expression recognizes the hierarchical subtext that can be smuggled into the concept of charity, while replacing it with a cherished Leftist category connoting strict equality.

It can be tempting to think of these material and immaterial offerings as gifts. For one thing, Anons may use the language of giving or, as later developed, "help." Nevertheless, in the anthropological canon, gifts have largely been thought of as an intervention into, or creation of, a social relationship, which through acts of transference sets up cycles of reciprocity. Indeed, as Marcel Mauss described, a gift may tacitly demand a counter-gift which, if not returned, can create asymmetries between the exchanging parties.[26] Although in practice, the underlying presence of transactions and asymmetries are hard to fully escape, it is significant that Anons energetically police the presence of both. By emphasizing that homeless people are "helping themselves" and attempting to avoid the ritual dynamics of handover, as well as their discursive and embodied efforts to flatten the interaction, these Anons seek to obviate the cycles of reciprocity that gifts produce, and the asymmetries that mount in the event of nonreturn. Instead, it is more precise to consider OpSafeWinter and its analogues as a form of sacrifice.

Sacrifice is derived from the Latin word *sacer*, meaning holy, and for most of the historical record has occurred within explicitly religious domains. Sacrifice can modally be said to occur when animals (or even humans) are ritually killed, or fruits and vegetables given up to rot, as an explicit act of communication with a deity or spirit world.[27] In these, the place or even time of the sacrificial act may be critical, occurring within specially bounded zones such as a temple or an altar, and/or at certain times of the day or year. In a recent discussion, Maya Mayblin and Magnus Course suggest that sacrifice retains an ethnographic salience beyond religion, and even beyond ritual, in many parts of contemporary life where it would not normally be recognized.[28] Instead of the historic emphasis on bloody ritual, they orient sacrifice around "the central idea that something (or someone) new can be created through the irreversible giving up of something else."[29] Even in nonreligious and nonritual contexts, Mayblin and Course hold onto the importance of an imagined realm beyond daily life into which sacrifice persistently reaches, where gods or spirits are substituted with other invisible beings who possess power over their lives, and with whom the sacrificer seeks to correspond.

The way Anons talk about and perform OpSafeWinter aligns it more closely with this long history of sacrifice than gift-giving. Every OpSafeWinter requires two things. The first is something that is being given up—coins, sandwiches, tins of food, even simply the boundedness of physical and emotional distance. The second is homeless people themselves. One Anon explains it in basic terms when he says, "If you have homeless in your town you have everything you need to start up." OpSafeWinter obviates the religious prescriptions of time and place, but as this utterance makes clear, in lieu of a spatial place it anchors itself to a human place, that is, the body of the urban homeless subject. Henri Hubert and Marcel Mauss, in their seminal study of sacrifice, assert that beyond a holy site, sacrificial killing is "mere murder"; and equally, without a homeless person, OpSafeWinter is mere loss.[30] For Anonymous, homeless people become moving altars, sites of communion that can have powerful sanctifying effects, yet they are also temporary and nigh impossible to return to. In the Facebook comments preceding this event in Nottingham, someone suggests enthusiastically that they try to find Steven, the man in the original film who becomes a digitally mediated archetype for the phenomenon. This is quickly dismissed as impractical, and like trying to find a "needle in a haystack."

Before developing the sacrificial character of OpSafeWinter further, it is worth drilling down into the regional history out of which it springs. Like initiation, sacrifice is an etic term, not one used by Anons themselves. Instead, when describing these activities, they are most likely to deploy the language of "help." The political antecedents of the word help in Britain provide a tool with which to excavate this history, and through it the wider forces of causation within which AnonUK flourished.

The Sacrificial State and the Logic of Austerity

Sociologist Richard Titmuss undertook a study of the remarkable rise in blood donations in Britain after the Second World War.[31] Between 1948 and 1967, the annual number of donations rose by 269%, an increase which could not be explained through material factors alone.[32] Why were Britons in the postwar era so prepared to give up part of their own vital substance for the benefit of strangers? To answer this question, he turned to a survey carried out in 1967, in which donors were asked to describe their motivations for giving. The majority of their responses fell under the category of altruism, of a desire to assist unknown others, many of the respondents referring to their own experience in the war and the importance of preserving life. What is striking is the ubiquity is of the word "help" across these responses, one that features in the majority of answers he cites in this category.[33] Titmuss applies the anthropology of gift-exchange to understand this increase, arguing that because it was anonymous, the blood constituted a "free gift"—the gift that demands no return because both giver and receiver are unknown to one another.[34]

Analytically, the idea of a free gift occupies the gray zone between gift and sacrifice, and Titmuss's material can also be unpacked through the latter. It is not only that the war is made explicit, and consequently along with it the blood sacrifice of those who died in the name of nation. The very substance being offered up here—blood—is (particularly in a Christocentric society such as this one) the ur-sacrifice, the original "something" that is given up for the sake of something else, underpinned by the mythic archetype of Christ's sacrifice on the cross.[35] The reason to dive into this postwar imaginary is to note the enduring idea of collective sacrifice in British national consciousness, one that is sometimes expressed through the language of help. Critically, this particular set of notions, of giving something up to serve a wider social-cum-national good, legitimized the creation of a National Health Service in 1948, to provide care free at the point of need, which then subsequently reinforced them through its own social and material infrastructures. As Titmuss says on this subject, "The ways in which society organises and structures its social institutions—and particularly its health and welfare systems—can encourage or discourage . . . generosity towards strangers."[36]

The moral logic of austerity drew with extraordinary efficacy on this cultural history of shared sacrifice. It was captured most pithily by Chancellor George Osborne's early assertion that "we are all in this together"—in which "we" are the citizens of the British nation-state, and "this" is a debt-burden requiring immediate payment.[37] Even if the national debt did require decisive action (although scholars soon pointed out that the more meaningful debt to GDP ratio was well within the normal levels), the unspoken reality at the heart of the austerity project was the gross inequality of the sacrifice itself, falling disproportionately on those who relied on the state for the exigencies of life.[38] In view of this, the fallaciousness of these words was immediately recognized by some, in one case being sampled in a music track that became a cult classic.[39] Yet it still possessed the kind of ideological force that resembled Margaret Thatcher's earlier claim that "there is no alternative," a phrase also resurrected to describe austerity.[40] Even though it was the subject of satire and derision by a left-wing minority, it still encapsulated the mass moral potency of the austerity project, which rode on this existing commitment to what Sarah Franklin calls "sacrificial reproduction," of giving something up for the promise of future endurance.[41]

One of the main architects of austerity, David Cameron, had been a member of the infamous Oxford University dining society, the Bullingdon Club. Although it has never formally been confirmed, it is a matter of public myth that the initiation ritual into the club involved immolating a fifty-pound note in the face of a homeless person.[42] In an act of potential mimicry in February 2017, a Cambridge University undergraduate filmed himself immolating a twenty-pound note before a homeless resident of the city. He shared it with a select group of friends on Snapchat but the video went viral, was picked up by the British media, and spread rapidly through Anonymous channels.[43] Anons were united in their horror and disgust at the act, prompting responses which ranged from signing a petition calling for his removal from the university, to veiled suggestions of vigilante justice. But the intensive focus on the student in question overlooks the wider and deeper set of values he was rehearsing. A member of the university's Conservative Association, his inebriated performance was the adolescent end of an ideological wedge of theories about the British public sphere that have been historically linked to its Conservative Party. In its spectacle of cruelty, this monetary immolation was also a sacrifice of a sort, a dark prayer which communicated that he possessed no duty of care toward the man in question. We might think of OpSafeWinter as its antithesis.

Eviction

While this OpSafeWinter is taking place, another drama is unfolding in Nottingham. Earlier that year, a 63-year-old retired flooring specialist, Tom Crawford, posted a plea for help on YouTube.[44] His story starts in 1988, when he took out a mortgage with a British building society in order to purchase his house. When, in 2011, the latter presented him with a repossession order for the property, he learned that the mortgage had been converted to interest-only payments without his knowing, and at the time when it should have been virtually eradicated, he still owed a significant sum. This instigated a three-year legal battle that was to culminate in the eviction of him and his family from their home. His plea to viewers was to come to his house on eviction day and physically prevent the bailiffs from entering, a form of direct action that has a significant history, particularly during economic downturns.[45] The convergence between the apparently simple injustice of the case, the accessible and lighthearted manner in which Crawford presented it online, and the wider context of critique of financial institutions that obtained in the wake of the banking crisis, made this an attractive political cause. The Pirates soon caught onto it and took the cause up with vigor, playing a visible part in the ultimately effective mobilization. Nonetheless, Crawford was subsequently issued with another eviction notice to take place in January 2015, just six weeks after OpSafeWinter. Still effervescing from the latter's enchantment, they prepare themselves for a different kind of intervention.

The morning of the eviction follows a night of subzero temperatures. Despite this the Pirates are outside Crawford's house before dawn, posting pictures of themselves under the orange glare of residential street lamps. The house itself happens to be strategically located, positioned at the end of a suburban cul-de-sac, on a raised area of ground only approachable by several steps. At the base of the steps the Pirates have staked their political territory, planting a large Anonymous flag that creases in the winter headwind at the top of a tall pole. When day breaks the number of people who answer Crawford's call for help on YouTube begins to rise steeply, consisting largely of friends and family of the Crawfords, a range of independent activists, some local eccentrics, and more and more men and women wearing Guy Fawkes masks. Indeed, Anonymous is by far the majority political affiliation on display, testament in part to the efficacy of its own call for solidarity with Crawford on YouTube.[46] Crawford himself seems rather taken aback by the degree of commitment the Anons exhibit. "They're absolutely brilliant," he says with a quiver of uncertainty, "I know they've got these faces, but my belief is that they use that for the people that are faceless."

As the crowd keeps growing, the atmosphere takes on the quality of an impending battle. Some Pirates employ direct military references, wearing camouflage clothing and advertising the need for "boots on the ground." At 10:45 in the morning, the now several-hundred-strong crowd finally meet their adversary in the form of the bailiffs, heavy-set men wearing large silver watches and dissimulating grins. Immediately, some Pirates position themselves at the vanguard, standing directly in front of the gleaming cars, phone cameras locked onto them like precision rifles. Because of the number of people who have come and the strategic location of the house, the bailiffs do not make even a modest attempt to repossess the property, and eventually the motorcade creeps away to victorious chants of "I am Tom Crawford!" and cheering. By all immediate metrics the action has been a resounding success, and in another video posted afterward, Crawford is shown giving a speech in which he extends a special thank you to Anonymous, again somewhat overwhelmed by the extent of their support. The conflict continues for months and years afterward, during which the Nottingham-based Anons continue to reinforce Crawford's moral right to remain in his home.

Meyer Fortes argues that the purpose of sacrifice is intrinsically defensive. It is a ritual encounter with the inescapable fact of human vulnerability in the face of far greater powers of life and death.[47] In the ongoing activism around the eviction, this defensive aspect becomes pronounced. Over a long period, Anons sacrifice significant time and energy, putting their own bodies on the line to defend the threshold of Crawford's house several times. At face value one might be tempted to read this as a Conservative defense of the rights of property (and this is likely why the drama is keenly reported in British right-wing newspapers), but in conversation with their nurturing efforts around OpSafeWinter, the Anons' quasi-militaristic defense of Crawford's house takes on a different kind of resonance.[48] For it is not simply his lawful right to property that is being defended here, it is in a concrete sense his home, a place with deep social meanings, where, as he says in the video, he raised his three children and buried his dead pets. It is now a good moment to reach into the imagined realm of sacrifice that Mayblin and Course identify. Why are homeless people—in one form or another—the altar on which Anons stake their resources? What exactly is it they are defending themselves against? Responding to these questions involves rehearsing the sequence of their initiations.

Pseudo-Incorporation

Sacrificial rituals are widely documented during the concluding phases of initiatory passage. They can take animal, vegetable, or mineral form, or some combination.[49] When they do, the sacred swap that happens during sacrifice—of destroying one thing to create another—becomes clear. A potsherd is smashed, a pig slaughtered, a wolf hunted, in the course of a larger process to make someone new, to seal the symbolic death of the initiand's prior self, and make way for a new state of moral being. Mayblin and Course, following Mauss and Hubert, stake the varieties of sacrifice they explore around this moral journey.[50] In the exclusively religious expressions studied by Mauss and Hubert, they privilege what they call expiation in this procedure, the banishing of evil forces that take place during the act of destruction, whatever form these take. But in a secular context like this, what is being expiated is not so clearly prescribed.

Anons begin their digital initiations with crises of separation, in which it is the felt absence of care that compounds the break. Critically for Anons, the failure of care becomes synonymous with unnerving sensations of invisibility, with not being considered of sufficient value to be an object of care. It is this wrong that is righted by OpSafeWinter. Anons often express their sense of sameness with homeless people. For Anons, the homeless symbolize these previous versions of themselves, and the piercing experience of symbolic death that sets their pilgrimages of digital knowledge in motion. This is pithily captured by the dominant tagline for OpSafeWinter, a phrase which accompanies the ritual online and off, "I learned to give not because I have much, but because I know exactly how it feels to have nothing." The sense of sameness is emphasized by this discourse of exactitude.

Being placed in this symbolic category can be bewildering for the homeless themselves. In the Steven video above, he is manifestly overwhelmed by the sudden display of solidarity, and it is the visible presence of affect, the materiality of tears, that gives the film its potency, as it visualizes the feelings that accompany these experiences of being invisible. But the homeless react in diverse ways. For some, the sudden shower of resources is a rare opportunity not to be missed, and they eke out each drop to maximize the potential material advantage. For others, what is being spontaneously laid at their feet seems almost excessive. One man lying in a doorway in central London, on the route of an Anonymous march, tells the tail end of the parade not to "bother"—he had already been presented with plenty to subsist on. Tom Crawford equally exhibits this bewilderment. He is grateful for their political support, but appears unsure of its real causation.

Victor Turner considers one of the defining characteristics of liminal personae to be their propertylessness. As he says,

> They *have* nothing. They have no status, property, insignia, secular clothing, rank, kinship position, nothing to demarcate them structurally from their fellows. Their condition is indeed the very prototype of sacred poverty.[51]

In building the homeless into the altar upon which sacrifices are made, there is thus a basic contradiction, at least with regard to Van Gennep and Turner's tripartite scheme.[52] OpSafeWinter is the sacrificial rite of incorporation that concludes their digital initiations, and yet it is one which still binds itself firmly to the practices and symbols of liminality.

In this regard it appears significant that during OpSafeWinter, Anons do not ritualize the sharing of food. There is one episode in Manchester that stands out. I get talking to two friendly homeless men wearing large rucksacks, who are waiting in line to be served hot dogs by a number of women wearing raised Guy Fawkes masks. They encourage me to take one too, though I am reticent, aware of the asymmetry of real need. Sure enough, when I arrive at the front, I am politely told to move on because I have "money in my pocket," confirming my misgivings. Although Anons emphasize their sense of sameness with homeless people, microsocial acts like these erect social differences between sacrificer and the subject of sacrifice, reinforcing the notion that there is symbolic work occurring. It also, in the process, distinguishes OpSafeWinter from many other incorporative rites that end initiatory sequences. These very often include commensality rituals, in which the sometimes ascetic period of transition is conclusively renounced, and new forms of social personhood intestinally recognized by the consubstantiality of eating together. This is one of the reasons why I consider these digital initiation rites as ending not in incorporation proper, but in a pseudo-incorporation, which mimics the idea of collective life, but without binding itself materially in the way that incorporation into enclosed communities does.

At the beginning of the book I suggested that online cults are characterized by being born digital, open-ended, with the ability to reach rapidly over large geographical scales, but potentially less ability to endure through time. The case of AnonUK suggests that the pay-off between openness and spatial reach, and enclosure and longevity, ultimately turns around the material practices and exchanges that the cult demonstrates the capacity to yield. Intriguingly, when parts of AnonUK do institutionalize the provision of food into an organization called Streets Kitchen (still operational in 2024, and in view of the current subsistence crisis more necessary than ever), it also ceases to be associated with the iconography of Anonymous.[53] Anonymous was born digital, and even in the most substantive incarnation of its ideals that we know of, in the final analysis remains largely so. Across the world, it continues to operate as a dynamic of unmasking, performing various kinds of question-begging wherever it appears. As the meme reads, while some people see it and think it's just a mask, others know it is a symbol for the idea of change.

CONCLUSION

A firework explodes on the back of my boot. The leather takes the hit, leaving only the heat, but when I see a long sear open like the eye of a lizard I know its days as a shoe are over.

Fieldnotes, 5 November 2015

The Million Mask March in 2015 is different from the previous year. The satirical and playful elements are still there—the jokes, the capes, the surreality—but they seem more out of place somehow, overshadowed by a more dominant seriousness of intent. The aesthetics of decoration have granulated and multiplied. Instead of the standard monochrome masks, many more people have arrived with their Guy Fawkes faces painted in diverse colors and designs, some going even further to coordinate them with the rest of their bodies—T-shirts, costumes, even tattoos. Alongside the usual consciousness and ovine referencing, other placards shout with a new self-assurance, "Knowledge is Power!" and "Knowledge is Free!" The previous year some participants had voiced concerns about a lack of noise. No such concerns are expressed now in my earshot. The popular chant, "One Solution, Revolution," is veritably bellowed as they make the move toward Parliament: shoulders back, strides long.

It is also much shorter. The march has followed a predictable route year after year, after Parliament comes Buckingham Palace, and London's Metropolitan Police have now strategized this eventuality. As the crowd turns away from Parliament Square toward the main thoroughfare that connects the two locations, rows of police are there to stop them. I clamber up onto a ledge for safety, and to survey the scene, and the vista that greets me is medieval. On one side are hundreds of activists, hatted, hooded, or crowned with masks, surging forward in great swells against the police on the ground, who thrash their batons at the vanguard. Several feet above the heads of the activists, a number of large flags suspended from tall poles ripple and bounce in the fray. The green and black flag of Anonymous is nowhere to be seen, but instead flies the symbology of Ano-nUKRadio, of the Pirates, of the County of Derbyshire, and the pagan figure of the Green Man among others. The lines of police on the ground, meanwhile, are reinforced by another line on horseback just behind them: a phalanx of infantry and cavalry guarding the route to the palace.

Although there are various flashpoints and flourishes after this, the confrontation is enough to suck any unifying momentum out of the march and leaves a feeling of unrealization among some of those present. A few days later, Tank tags me and others in a poem he has written about the night.

> Gunpowder deliveries first right
> more than just a joke.
> Feelings high on that night
> millions of us spoke.
> Anonymous is changing
> more than just a mask.
> Police disrupt, some shame us
> undercover task?
> Bad weather and their planning
> have made this year's damp squib.
> A stage in our evolving
> move on from ad-lib.

In observable ways, 2015 marks the beginning of the end for AnonUK as an online cult. The communitas that had infused the previous year is not there to the same degree. Old walls of difference are restored. Participants are less willing to be interviewed, more suspicious of being filmed, and while it is pleasing to reconnect with all the people I know, for those I do not my identity as a researcher is reoriented toward Otherness. On sensing the absence of communitas, it retrospectively reveals the strength thereof that AnonUK had previously been able to nourish. One of the original progenitors of the march articulates the same sense of absence, posting online shortly after it is all over, "WHAT HAPPENED TO THE MAGIC!!!" His question reinforces Taussig's ideas about the magic of unmasking, particularly its fleeting quality. Perhaps by this point it had all been released.

While AnonUK gradually ceases to exist as a socially active force in the months and years that follow, as continuing historical actors, those who have been digitally initiated into it do not. To reflect further with this view of hindsight, let us then return to the three protagonists of this process whom I meet again in the summer of 2017 and explore the shores onto which AnonUK has washed them.

The Fatalist

Tank suggests we meet outside a gallery in London, and on arrival he appears just the same. Same crew-cut hair, same style of polo shirt, same sense of humor which has me frequently collapsing in hysterics. And as he talks, familiar topics resurface: his father's ventriloquism of the *Daily Mail*; his socially mobile upbringing; the materialism and binges that characterized his life as a young man. But there is a new ease in his words and manner, one that suggests a new orientation toward these topics. He is more forgiving of his father's fixedness, more accepting that commodity culture is a source of meaning for some. He has recently started a new job as a business development manager for an IT company, which he tentatively says he enjoys. He has also since become teetotal, and he no longer spends hours every day on Facebook.

As the conversation unfurls, it becomes apparent that this new ease has come about through his own near complete detachment from political desire. He stresses at several points that he has come to realize that he cannot change the world. Tank talks in pictures, and at one point paints the following:

> I think our society is a bit like a cup of water, and you put your finger in it, and then the change that is left is the hole that is left when you take it out

This is his way of saying that no such change occurs. What distinguishes Tank from activists in other movements, is that this feeling of a lack of agency is not simply a temporary burnout but takes on a rationality of its own. Though he does preface the statement by saying, "This is going to sound really terrible," he now believes there is a "natural order" to this stasis. Recalling Eliade's theory of nonmodern temporality, in which time is conceived not as a straight line but as a circle that repeatedly returns to the start, he now thinks of social change as cyclical.

> It's just like a cycle. And you have all these little things that happen, and the bigger picture is we've shifted around slightly

Change in this formulation takes on a quasi-cosmological aspect, like orbiting planets. What is consistent with his prior self, however, is that the change is still ultimately wrought by knowledge. His feeling of the futility of activism is prompted, ultimately, by the enduring "ignorance of the masses" in spite of Anonymous's endeavors at awakening.

One might guess that the consequence of all this would be a new perspective on the knowledge he gained on waking up, but Tank is adamant this is not the case. He "absolutely" still believes in the Illuminati and its nefarious project of depopulation. As he describes it,

> The problem is that once you are enlightened, once you see what's behind the curtain, it's very hard to put Jack back in the box

His new ease has arisen, not through the reconfiguration of what he knows, but how he relates to it. Indeed, when Tank describes knowledge, what precisely he saw behind that curtain, it takes on an object-like quality, like a piece of furniture. The burden of it has been removed because he has taken it off his back and put it "over there." It is still there, he says, but he does not wish to focus on it any more, as there is more to his life now.

The Faithful Anon

Patrick went through one of the most profound digital initiation rites recorded, involving a visionary experience involving the Guy Fawkes mask. On the other side, he remains among the most loyal. Patrick describes his relationship to Anonymous using the acronym AWOL, standing for Anonymous Way Of Life. When pressed he defines it in the following way:

> A rebirth . . . Anonymous gives you a way of going, right, you might have done wrong, but now you can see that the system is rigged up for you to do wrong, and you have seen past that and you awaken . . . So now you don't really want to fight that person over there, you don't want to take hard drugs, because your brain is working. Anonymous has opened your brain, awakened it, to start looking at life.

To the question of what fills the space vacated by his former self he replies instantly, "Knowledge. Knowledge and truth." This is what has elicited morality and given him purpose. He always knew a lot, he says, even before the Internet existed, but it was this medium that allowed him to piece the parts together, to assemble it all. The subject of why AnonUK has diminished over the previous two years preoccupies him, and he returns to it repeatedly, concluding that it was eaten from within by infiltrators and double agents. In spite of its decline, he still sees Anonymous as a major generator of historical change, and deploying the favored portal metaphors, says it has "opened up a door" for all kinds of other activism since. At one point he remarks (with perspicacity given I had never described myself in these terms), "You're our historian." It is a statement that carries a sense of the cult as past, as well as the sense of grandeur that had characterized many people's experience of it.

As with Tank, however, the extreme agency that Anonymous was able to inspire has evolved into an equally extreme form of fatalism. Unlike Tank, Patrick is unapologetic about his theories of a natural order, which involve the movement of the earth and the planets, and the misty assembly of the New World Order who are now preparing for their final reckoning. Patrick makes links across our discussions, reminding me how everything he had predicted in a previous discussion is now occurring, a process which continues right up to the coronavirus pandemic of 2020. Though his explanations are abstract, ultimately Patrick is seeking to explain the concrete conditions around him, like the pandemic, or the disappearance of skilled jobs in his city. Moreover, one of his prophecies does come true two years after our final meeting—that Boris Johnson would be Britain's next prime minister—a proposition which at the time I balk at. The concurrence of this fatalistic reasoning with the actual demise of AnonUK seems to propel Patrick into a state of waiting, one that may be exacerbated by his physical deterioration. He is now on medication for his nervous system that has made him put on weight, and having formerly been energetic, he now carries a cane to help him walk. Juxtaposed with this frailty, though, is the hooded sweatshirt he wears, with a list of raves that took place across England in the 1990s on the back, a memory of communitas to which, like Anonymous, he remains cathected.

The Trump Supporter

Betty collects me from the station in her immaculate car, and she is looking as glamorous as always. She has dyed her hair from platinum blonde to a coppery color, and she tucks it neatly beneath a black flat cap, all set off by a keffiyeh draped artfully around her neck. Like last time I was at her house the day is sunny, and she suggests we sit in the garden. There are clumps of sage, rosemary, and potatoes growing, as well as a pot with some spring onions in it, which she replanted using a cuttings technique she saw online. She has become even more enthusiastic about alternative remedies and expounds on the healing properties of onions, turmeric, and other plants. This time, however, she does not offer any cooled water, as she tells me she no longer drinks it, and we slurp cups of strong milky tea instead. When we later relocate to her living room, the single Anonymous mask that was on display has been joined by seven others—six of which are painted in different colors and styles, and one which remains in its cellophane wrapper, intended for the next visitor who expresses an interest in Anonymous. The other noticeable change is the presence of a small screen next to them, which shows rolling footage from the surveillance cameras she has recently installed outside her home.

In contrast with Tank, Betty has not filed her knowledge away, but every day goes deeper into it. Each morning, she says, she makes a cup of tea, rolls a cigarette, and goes straight to YouTube. "I am constantly researching and reading about things and looking at films," she says. Betty rehearses some of the theories she had described before, but now they are more elaborate, more extensive, and for her, more coherent—"everything is slotting into place," she says. She supports erstwhile President Trump (who she describes as "astute"), as well as Britain's Nigel Farage (she voted to leave the European Union). Yet Betty's political position is not easy to pin to one end of the political spectrum. She combines the backing of right-wing demagogues with radical left-wing commitments—as someone who is vociferously antiracist, anti-imperialist, and anti-Islamophobic. She takes exception to Trump's border policies because America is Indigenous people's land anyway. Betty is observant and alert, eyeing my reactions closely when she speaks. At one point she catches me off guard, when she fires a volley of questions back at me.

> Would you rather know or not know? Would you rather be a sheep or would you rather be awake? Which?
> Awake? (*I reply uncertainly*)
> But *would* you? Because it's so dark V.

There is a yearning in her voice as she says it; this is a genuine question for her. Online, she has discovered whole new communities of people she can talk to, people who know what she knows, but in-person she says, socializing has become more difficult than it used to be. People laugh in her face, she says, and there is now a much smaller number of people she can truly converse with. As Betty returns throughout to the theories that govern her cognitive world, they are not only more elaborate and all-embracing, they are also palpably more extreme. She now denies the existence of climate change. The reason she no longer drinks water is because she believes it is being poisoned. In view of Betty's long biography, this takes on a particular poignancy three years later in the midst of the Covid-19 pandemic. Despite the fact that her life course was shaped by catching polio just before its mass vaccination programs, Betty is unwavering in her refusal to be vaccinated for Covid.

Knowledge and Social Change

Each of these protagonists indexes the permanence of the change made by initiation. Having acquired knowledge online in crises of separation, there is no going back to how they were beforehand, and this is the case for all key participants.

Their relationship to the knowledge may change, manifested in the contrast between Tank's archival approach to what he learned, versus Patrick's and Betty's orientation of it toward the future, but the knowledge itself does not. This capacity for permanence is described by Jean La Fontaine:

> Initiation rituals are almost always irreversible; an insider cannot reverse the process and become an outsider again. Knowledge once acquired cannot be unlearned.[1]

She describes this phenomenon with particular reference to a secret society in Kenya known as the Mau Mau, a guerrilla organization that opposed the British colonial authorities in the 1940s and 50s. Entry to the Mau Mau involved taking strict oaths of silence. La Fontaine observes that even despite the fact that some oaths were taken in conditions of coercion, counter-rituals devised by the new Kenyan government to release Mau Mau members from their oaths proved overwhelmingly unsuccessful. This was attributed by both opponents and supporters alike to the power of its initiation rites.[2]

Those invested in the cyberutopianism that presaged the rise of Anonymous sometimes express perplexity that the world they envisaged did not come about.[3] Why did the "freeing" of knowledge not ultimately topple the old power structures that emerged in conditions of its constraint? To this question I submit the response, that what cyberutopians had not prepared themselves for, was the deeply asymmetrical character of knowledge itself. Knowing something means bearing something that someone else does not bear, and the difference is not an equal one. In traditional initiation rites this asymmetry is acknowledged in bluntly hierarchical terms, where the difference between the initiated and the uninitiated turns around the powers of sight that arise through the presence of knowledge. This is also the case for Anons, whose triangular world falls partly along fault lines of knowledge and nescience. Nonetheless, AnonUK was a tiny minority within a much larger majority society that still subordinates the knowledge they have acquired through their digital initiations, to the knowledge produced with and through educational institutions. The very category of conspiracy theory can be understood as an outcome of this continual hierarchizing process.

Naturally this stimulates reflexive questions. Is this not also happening here? Is it not my own three decades-long passage through educational institutions that places me in the position to pen this ethnography, and to present the world of AnonUK as something other from my own? At least methodologically, this is true. Anons were aware that I did not share their theories of depopulation, neither its mechanisms nor its authors, which is partly why some of their relations toward me are evangelizing. And yet I remain discomforted by the language of conspiracy theorists, because it displaces the hierarchizing of knowledge onto the hierarchizing of human beings. It is its own peculiar power play, whose effect is to delegitimize their subjective testimonies in their entirety. I employ the concept of vernacular theory to obviate the latter and—without endorsing their theoretical positions—dignify the embodied experiences they have undergone. As in other political domains, for a progressive politics to succeed, it must be possible to disentangle *theory* from *theorists*, to allow for the possibility of theoretical change.

What then, are the broader methodological and social implications of digital initiation rites? This largely remains to be seen, but here I offer some preliminary reflections.

This book has sought to contribute to the discipline of anthropology, by considering digital initiation as a novel social phenomenon, as a ritual "in its own right."[4] Here I have been influenced by propositions of Van Gennep and the later Manchester School, particularly Victor Turner's assertions about the transformative effects of liminal experience. At the same time, I have traveled with Don Handelman's critique thereof. For Handelman, the issue that has bedeviled the study of ritual has been that very different social phenomena, with very different social outcomes, have been considered within the same overarching category. More specifically, he proposes, there are rituals that are continuous with the social order and whose purpose is to represent it, and there are rituals that are autonomous from this order, and hold the capacity to transform it. The difference between them is located in their own interior complexity. A digital initiation rite certainly falls closer on the side of the spectrum of rituals that are autonomous and transformative, rather than continuous and representative. As a phenomenon it shows, not only, as Miller and Horst suggest, that digitalization is changing the nature of the human, but also that it is transforming *how the human is changing.*[5]

That digital initiation rites indeed possess an extraordinary degree of interior complexity, far more indeed than it may be possible to document with ethnographic methods alone, suggests its capacities for social transformation are profound. Indeed, there are indicators that Britain has entered a new era of, as Helen Margetts and coauthors put it, "political turbulence," as a consequence of digitalization, where large sinkholes in the social landscape suddenly appear as if from nowhere.[6] Kenelm Burridge argues that innocuous cults create the conditions for more violent cults to emerge.[7] The rise of QAnon in Britain might be seen in this light. Elsewhere, during the antimigrant Islamophobic riots that erupted in the summer of 2024, photographs showed some participants wearing Guy Fawkes masks.[8] While anthropologically distinct from my own interlocutors, it seems plausible that those wearing them had undergone parallel transformative processes.

This ethnography has been a work of description and analysis, of assembling complexity, not a political program nor policy proposal. However, if I were asked to suggest a programmatic response to this research, it would be the following. In contrast to the cognitive orientation of much of the writing that seeks to explain the rise of online extremism, the body continues to be pivotal to the change that is wrought by the knowledge that arrives online. The transformation of waking up is prefigured by varieties of structural carelessness, sometimes extending over many years, even decades, suggesting direct links between existential care at the time of acute need, and the shape of political loyalties thereafter. These links are something that members of cults (and some social movements) have long intuitively grasped: that the deepest needs of human beings are for forms of embodied and existential care, and that if these needs can be met, more abstract loyalties will necessarily follow. The political actor is not (and has never been) a rational actor.[9] The implication, then, would be to hold firmly onto the structures of care that already do exist, embedding the theoretical reasoning for their own reproduction into them, while continuously seeking to build new ones. Not through the hierarchical dynamics of charity (i.e., the food bank), one might add, but as the highest expression of social solidarity.

Acknowledgments

A labor of wonder over the span of twelve years, this book has incurred many debts. My first acknowledgment goes to all the Anons who were willing to share their lives with me, without whom this book could not have come into being. While the themes you raised are serious, fieldwork was largely a light and joyful experience, and the laughter we shared bubbles between the lines above.

This project was seeded and researched in the rich intellectual environment of University College London. I would like to thank the Material Culture faculty for early instruction and Charles Stewart for graceful mentorship. Conversations across the department informed the research, but I'd particularly like to thank Allen Abramson, Ludovic Coupaye, and Jerome Lewis, as well as attendees of the *War of Worlds* workshop, Angelique Haugerud, Alex Flynn, Stine Krøijer, Ellen Potts, Maple Razsa, and Jarrett Zigon. David Jeevendrampillai, David Lynch, Lucas Somavilla Croxatto, and Rebecca Reid joined me on some of the Million Mask Marches and contributed to the wealth of fieldwork through photographs, transcripts, and valuable reflections. I am sincerely grateful to the Economic and Social Sciences Research Council for kickstarting the project with a Future Research Leaders Fellowship between 2013 and 2016, and to Georgie Aronin for hundreds of pages of meticulous interview transcription.

In autumn 2015 I spent a magical semester at Columbia University New York (and amid the thinking fields of Manhattan provided by NYU and CUNY respectively), where some of the main fault lines were established. I am especially grateful to Mick Taussig for generous and engaging hosting, including unforgettable seminars on the Art of Fieldwork. Danielle Carr, Gustav Kalm, Rune Steenberg, and Wilson Villaverde all offered insights into the research during this theoretically formative period, as well as the delights of friendship.

In the years thereafter, conversations in Cambridge have infused the work. I would particularly like to thank Sian Lazar and Piers Vitebsky, who in different ways headed me off the anthropology of movements and stimulated thinking around cults. Lively discussions with Christopher Clark, Anita Herle, Shruti Kapila, and Simon Schaffer have always been founts of energy. Critical readings of some chapters were generously afforded by Natalie Morningstar, Kelly Robinson, and Fiona Wright, who helped me to separate the wood from the trees and view the argument from different angles. Joel Robbins has been a particularly thoughtful interlocutor, inviting me to consider the work of Don Handelman; while Richard Irvine, Michal Murawski, Jonas Tinius, and Khadija von Zinnenburg Carroll have all in different ways illuminated the research. I'd also like to thank the staff of Cambridge University Library for stewarding its extraordinary resources.

I offer warmest thanks to Dominic Boyer, Jim Lance, Bethany Wasik, Susan Specter, Lucinda Treadwell, and the whole team at Cornell University Press, for showing me what excellence in publishing looks like, and infusing the experience with generosity of spirit. Lisa DeBoer provided a superb index. Thank you to Deniz Yonucu for pointing me in this direction. I am also grateful to the editors of *Surveillance & Society* journal, for allowing the restatement of parts of the argument in chapter 3 (cf. Peacock 2025). Anonymous reviewers of this book also helped me to collect and advance its main assertions, and the European Research Council (grant no. 947867) provided welcome support while this book was in production.

Written and rewritten by hand over many years, and through the undulations of the early twenty-first-century scholarly life, my final acknowledgment goes to all my dear friends and family who have been present at every stage. Of special mention are my parents, Andy and Jila Peacock; my brother and sister, Johnnie and Leila; Richard McKay, who cast a keen eye over part of the manuscript at a crucial moment; and Benjamin Elwyn, who skillfully edited some of the images. Over this period I myself underwent two rites of passage, marrying my partner, Richard Drayton, and being blessed by my daughter, Daria, and as I read the final proofs, my son, Jasper. I'm thankful to Claudia Cardoso and the staff at Queens College Day Nursery for being important parts of the village that raised Daria to allow the writing of this book, and to Richard for taking her on long walks across Grantchester Meadows. My deepest acknowledgment goes to Richard, to whom this book is dedicated, and whose love, patience, and daily inspiration are the nest in which it grew its wings.

Glossary

Depopulation Agenda Theory that a globally connected group of powerful actors seek to reduce the earth's human population to five hundred million

digital initiation rite Ritual sequence of biographical change comparable to an initiation rite, in which digital communication and information technologies play a substantive role in part of the sequence

initiand Person undergoing initiation

initiation rite A ritual sequence of symbolic death and rebirth, charging the initiand with knowledge about a society that produces a moral responsibility for that society

liminality Transitional phase during a rite of passage in which the subject becomes structurally invisible (also known as the "liminal period")

OpSafeWinter A ritual performed at any time of year, either by an individual or by a group, in which donations of money or needed items are distributed to people identified as homeless

symbolic death Temporary condition of symbolic invisibility experienced by the initiand in the liminal period, rendered through a variety of methods (coined by Victor Turner [1967])

thanatic mask Mask deployed as part of a ritual sequence to reveal the face of the wearer

thanatoid mask Metaphor of unmasking deployed visually or verbally

They Minority group of powerful people who are waging a hidden war on We/You, sometimes expressed as the Depopulation Agenda

waking up Process of subjective transformation which turns on the acquisition of political knowledge

We Minority group of people who are awake, that is, know of the hidden war They are waging on We/You

You Majority group of people who are asleep, that is, do not know of the hidden war They are waging on We/You (also known as "sheep," "sheeple," or "sleeple/sleople")

Notes

INTRODUCTION

1. Gilbert and Driver 2000, 29.

2. The "triumphal axis" was designed in 1913 (Gilbert and Driver 2000).

3. Moore 2008.

4. This video is no longer publicly available. The original URL is: https://www.youtube.com/watch?v=OQA_JwD4dOs.

5. A note on quotations. Where a word or phrase is within quotation marks but not referenced to a primary or secondary source, this has been cited from ethnographic interviews. All interviews were conducted in confidentiality, and the names of the interviewees are withheld by mutual agreement. More detail on methodology can be found on pages 16–18.

6. It is not known who first renamed #OpVendetta the Million Mask March, though the construction and the protest history it references derive from the United States. In 1995, more than a million African American men marched on the Capitol building in Washington to raise awareness of enduring racial inequalities, in what was called the Million Man March. In 2004, the satirical activist group Billionaires for Bush also participated in an ironic Million Billionaire March to protest against rising inequality (Haugerud 2013, 15).

7. The term *cult* remained a largely value-neutral category in social science—i.e., one employed to make an analytical rather than a moral point—until around the 1970s. The use of the term—for instance in "mystery cult"—is still value neutrally applied in classical studies.

8. On this folk history see Cressy 1989; 1992.

9. See Cressy 1992 and Hutton 1996 on practices of fence-stealing in the eighteenth and nineteenth centuries. On the latter, Hutton (1996, 399) documents that, "On 5 November 1828 a masked crowd threw stones and fireworks at the house of an unpopular employer at Luton."

10. Cited in Hutton 1996, 382. The "Bonfire Boys" remain an enduring institution in Britain's largest bonfire night festival in the town of Lewes, although much of former rule-breaking character of the Bonfire Boys has been contained through its institutionalization (Etherington 2001).

11. See Hutton 1996 for a thorough examination of this festival on the British ritual calendar.

12. Eliade (1955, 53–70) notes the preponderance of initiation ceremonies during seasonal rites among the Native American Hopi and across Indo-European societies. Reinforcing Eliade's analysis, Pierre Smith (1982) documents from his own ethnography among the Bedik of Senegal, how the first stage in the initiation of young boys into adulthood took place within the framework of an annual festival that was otherwise distinct from it.

13. This is a police tactic to surround demonstrators in order to contain them in a particular place. It was first used in Britain in the late 1990s but became particularly widely used in the post-2008 period (cf. Neal, Optiz, and Zebrowski 2019).

14. As "anti-capitalist" see *Daily Mail* (Cockcroft 2014), *The Mirror* (Grierson 2014), *The Telegraph* (Harley 2014), *The Independent* (Saul 2014), *The Guardian* (Quinn 2014); as

"anti-establishment" see BBC News (2014a); *Daily Mail* (Cockcroft 2014), *The Telegraph* (Brooks-Pollock 2014; Harley 2014), *The Mirror* (Grierson 2014); citing Anonymous see *The Mirror* (Grierson 2014) and *The Standard* (Rucki and Morgan 2014).

15. Although my interlocutors would not use this designation, it is a popular Twitter handle to reference Anonymous in Britain and is also contained within the name of the influential radio broadcasting collective of the period, AnonUKRadio.

16. Beyer 2014; Call 2008; Coleman 2014; Deseriis 2015; Firer-Blaess 2016.

17. Miller and Slater 2003; Horst and Miller 2013; Hine 2015; Pink et al. 2016; Geismar and Knox 2021.

18. Miller and Slater 2003, 5.

19. Turner 1969, 95.

20. Van Gennep 1977.

21. Max Gluckman (1962, 2) calls this his "pioneering discovery."

22. Van Gennep 1977, 21, 36.

23. Van Gennep 1977, 21.

24. Van Gennep 1977, 21.

25. Turner 1967, 93.

26. Turner 1974a, 273.

27. Turner 1974b.

28. Turner 1978, 287. In a trenchant review essay, Don Handelman (1993) argues that some of the work Turner posthumously inspired did indeed possess the kind of speciousness the latter feared.

29. Handelman 2021, 63–92.

30. Handelman 2021, 67, 71.

31. Van Gennep's (1977) chapter on initiation is fifty-one pages long, while the next most extensive chapter on betrothal and marriage is thirty pages. In *The Ritual Process* Turner's (1969) discussion of liminality is oriented around an installation rite and a puberty rite. His celebrated essay on liminality also turns on these two varieties of initiation.

32. Turner 1967, 95.

33. Liminality is "a realm of pure possibility where novel configurations of ideas and relations may arise" (Turner 1967, 97); something which "generates imagery and philosophical ideas" (1969, 133); a "gap between ordered worlds (where) almost anything may happen" (1974a, 13). Turner (1974a, 17) at one point acknowledges his basic assumption that he "regard(s) mankind as one in essence . . . creative."

34. "All sustained manifestations of communitas must appear as dangerous and anarchical and have to be hedged in with prescriptions, prohibitions, and conditions" (Turner 1969, 109).

35. "The life of an individual in any society is a series of passages from one age to another and from one occupation to another . . . such acts are enveloped in ceremonies . . . so that society as a whole will suffer no discomfort or injury" (Van Gennep 1977, 2–3).

36. Handelman 1990, 66.

37. An Internet Relay Chatroom or IRC is an enclosed virtual space where messages are exchanged between members, the precursor to the direct messaging functionalities on today's social media.

38. Parmy Olson (2013, 28) writes that a program called "Forced_Anon" was implemented on several of 4chan's messaging boards at the behest of Christopher Poole, its founder and main administrator.

39. Firer-Blaess (2016, 162) describes the deviance that attaches to naming during these years as an "ethics of self-effacement," in which users who identified themselves in or outside the forum were roundly mocked.

40. Firer-Blaess 2016, 116.

41. Coleman 2014, 33–51. On the association between this era of Anonymous and play see also Deseriis 2015, 166–73.

42. Firer-Blaess 2016, 125–26.

43. On Free Culture see Swartz 2016 and Jimenéz and Estalella 2023.

44. See Juris 2008 for an ethnography of how these ideas flowered in the anticorporate globalization movement.

45. Cited in Castells 2012, 159.

46. On 15-M see Castells 2012, 260–70; Postill 2014.

47. See Coleman 2014, 143–76.

48. See Castells 2012, 274–75).

49. Beyer 2014; Coleman 2014; Firer-Blaess 2016; Olson 2013. Coleman (2014, 175) suggests that many of her participants were "generally male, middle-class, libertarian," and of those she came to know personally, the majority were under thirty. Beyer (2014, 148–49) corroborates the likelihood of first wave Anons inhabiting this particular age bracket, through several external data indicators.

50. Firer-Blaess (pers. comm.) cites personal risks on the side of the interviewee as impacting his methodology. Beyer (2014, 157) describes her impersonal approach as "low-risk," in which the "researcher is interested in profiling aggregate behaviours not telling individual stories." Coleman (2014, 13–16) intimates concerns about endangering interlocutors when visiting the Canadian intelligence service.

51. I had prepared for fieldwork by learning a formal shorthand system, to record interviews by hand in anticipation that my interlocutors would not wish to be digitally recorded. By contrast, all key participants were willing to be recorded. This was one of many ways in which AnonUK stood at odds with the ethic of anonymity.

52. It should be added that participation in the fieldwork was to a degree self-selecting. The biographies and activities of those chronicled below were people who sought out documentation. The small minority of those in AnonUK who kept their faces covered, still guided by an earlier ethic of anonymity, are present here as voices but not as characters.

53. Büscher and Urry 2009, 110.

54. Hine 2015.

55. With respect to data ethics, research material was gathered from social media where this was intended for an entirely public audience, in other words commensurate with acts of public speech during the in-person events I recorded and transcribed in field-notes. Where this material fell under any privacy setting it was not included in the data set.

56. Of these, forty-four were carried out by David Jeevendrampillai, Rebecca Reid, and Lucas Somavilla Croxatto with attendees of the Million Mask March in 2014 and 2015. The research generated an ethnographic data set of 720 single-spaced pages of fieldnotes and interview transcripts, approximately 2,000 photographs, videos, pictures, and screenshots, and five handwritten fieldwork diaries. Some redacted long-form interviews are available to read by registered users of the UK Data Service (http://doi.org/10.5255/UKDA-SN-852720).

57. Postill 2012.

58. Deseriis 2015.

59. Blyth 2015.

60. For a comprehensive account of the 2008 financial crisis and its aftermath see Tooze 2019.

61. Duffy 2013, 10.

62. For some of the structural effects of this policy choice see Atkinson, Roberts, and Savage 2012 and O'Hara 2015. See Duffy 2013 for a table of government spending changes. The NHS budget was not cut during this period, but in a context of rising costs and inflation, this amounted to a reduction in real terms.

63. Duffy 2013, 12.

64. A key function of British local government is to provide social care; hence the disproportionate reduction of local government spending disproportionately affected the social care system.

65. Duffy 2013.

66. The number of users of Britain's biggest food bank grew from 41,000 in 2009–10, to 1.1 million in 2014–15, with the most frequent explanation for need bearing some relation to the benefits system (Butler 2015). In 2017, a group of health economists tracking mortality data showed that austerity policies had produced an extra 120,000 deaths in England alone, most attributed to a shortage of nurses (Watkins et al. 2017).

67. Atkinson, Roberts, and Savage 2012.

68. Winston Churchill's 1929 Budget Speech cited in Konzelmann 2019, 62; see also Blyth 2015.

69. Cited in Konzelmann 2019, 117.

70. Rakopoulos 2018.

71. Konzelmann 2019.

72. Kalb 2018, 142.

73. Weber 1966; Troeltsch 1981. See Dawson 1998 for an overview of the sociology of cults.

74. See Burridge 1960, 1969; Jarvie 1967; Lawrence 1964; Mead 1956; Wallace 1956; Worsley 1957.

75. See Barker 1984; Robbins 1988; Stark and Bainbridge 1985.

76. Stark and Bainbridge 1985, 73.

77. On twentieth-century cults as hierarchical see Tourish and Wohlforth 2000.

78. Diefenbach and Sillince 2011; Freeman 2013.

79. Alberoni 1984, 34.

80. Latin for "community," communitas is a concept deployed by Victor Turner to describe the social bonds that characterize the time of liminality. A deeper discussion of this concept can be found in chapter 4.

81. Dawson 1998. The massacre of 918 people in Jonestown, Guyana in 1978 by the "People's Temple" cult constitutes a major turning point in the scholarly and popular consideration of cults.

82. In Van Badham's (2021) account of QAnon; these two terms come together as a "conspiracy cult."

83. David Robertson (2016, 39), in his ethnography of comparable theorists, takes a similar stance, saying he "cannot use the term 'conspiracy theory' in good conscience."

84. See Birchall and Knight 2022.

85. For more detailed chronologies of QAnon see Ball 2023 and Van Badham 2021.

86. Marwick and Lewis 2017, 18. See also Anon 2021; Ball 2023; Nagle 2017.

87. See Coleman and Golub 2008; Deseriis 2015.

88. Udupa 2019.

89. Aly et. al 2016; Littler and Lee 2020; Kruglova 2022.

90. Conway 2017.

91. See also Munn 2023.

92. On the resurgence of comparison in anthropological work see Candea 2019 and Van der Veer 2016.

93. Note on the text: each chapter begins with a thick description which, while presenting the theme to follow, also collapses differences between traditional initiation and some of the audiovisual artifacts that bind AnonUK.

94. cf. Horst and Miller 2020.

95. Cited in La Fontaine 1985, 20.

96. Eliade 1995, ix.

97. On the advance of evangelical Christianity replacing traditional initiation in Papua New Guinea see Robbins 2004; on the spread of Islam in West Africa see Mark 1992.

98. cf. Herle and Philp 2020.

99. Myers 2017, 10.

100. This also resonates with the approach emphasized by Sanchez 2023, of renegotiating rather than ending relationships with disciplinary ancestors.

101. On the continuation of feudal kinship relations in this region for more than a millennium, see Clark 2014.

1. BODY

1. Cited in Bloch 1992, 9. This narrative is retold from Bloch's account, which is itself a rehearsal of André Iteanu's study of the ritual.

2. In his summary of the abundance of death imagery, Turner (1967, 96) says, "The neophyte may be buried, forced to lie motionless in the posture and direction of customary burial, may be stained black, or may be forced to live for a while in the company of masked and monstrous mummers representing, *inter alia*, the dead."

3. The clipping or shaving of the hair in a practice known as "tonsure" (*Britannica* 2024), is found in initiation rites in a range of religious orders, including Buddhism, Jainism, Hinduism, Eastern Orthodox Christianity, and Roman Catholicism, in some cases forming the central occasion of the rite. Patterson (1982, 60) notes the ubiquity of head shaving in rites of enslavement. As Turner (1967) observes, many initiands are covered in black to evoke their new corpse-like state, such as in Northwest Amazonia where boys are painted black "from toes to chin" (Hugh-Jones 1979, 68). But white may equally be used to evoke the symbolic death of the initiand, for example in male puberty rites on the Loango coast of the Congo (Eliade 1995, 31), and in the puberty rite of the Gisu of Uganda, where boys are smothered in a white yeast paste (La Fontaine 1985, 121). A particularly vivid example of coverings as a rite of separation occurs among the Bemba of Zambia, where the puberty ritual begins when the girls crawl into the initiation hut covered in blankets, described in chapter 5 (Richards 1988, 121). On the practice of blindfolding see chapter 4, note 38.

4. On male initiation huts see Barth 1975; Hugh-Jones 1979; and Taussig 1999. On female menstrual lodgings see Lincoln 1981. On masonic lodges see Wilmshurst 1924 and Runton 1942. Van Gennep (1977, 85–108) describes the Christian ritual of penance which requires withdrawal into a monastery, as well as wandering among the Masai and woodland isolation by Australian Aboriginal shamans.

5. On all of these mutilations see Van Gennep (1977, 70–109). The extraction of teeth, pulling out of hair, tattooing, and lacerating of the back are further documented in Australian Aboriginal societies (Eliade 1995, 4).

6. The practice, known as *Mensur,* is described by Norbert Elias (1996). The scars were common markers of belonging across upper-class German society over this period, displayed by many prominent figures of the time such as Max Weber.

7. Marwick and Lewis 2017; Munn 2023.

8. Marwick and Lewis 2017.

9. Kruglova 2022.

10. Ball 2023.

11. Marwick and Lewis 2017; Conway 2016; Kruglova 2022.

12. Social Services entails the range of child protection agencies which are part of the UK government. A case worker is a social worker employed to provide guidance or information to vulnerable citizens.

13. Barker 1984.

14. The text is named *The Divine Principle* and was the central source of doctrine for the Unification Church.

15. See Burridge 1969.

16. Certain cis-masculine norms come through in Patrick's account that demonstrate the relation he sees between gender and power.

17. Barker (1984, 218) states that three quarters of those who joined the Church reported having visions beforehand, some of which involved the presence of God or Jesus speaking to them.

18. The idea of social change as an incremental process, originating in small-scale acts by individuals, is addressed in chapter 7.

19. The video is titled *Omnipotent* and is a rap collaboration by Eminem, Ice Cube, and Korn calling for a "worldwide wave of action" at sites formerly inhabited by Occupy. This crowd-sourced protest took place on 4 April 2014.

20. A Local Authority is a subdivision of the UK government that is responsible for local services, such as Social Care, within a given area.

21. See Introduction for further details.

22. A sanction is the withdrawal of a welfare payment by the Department for Work and Pensions, if the claimant has failed to fulfill one or more obligations. Conditionality around the receipt of welfare payments was expanded under all three major political parties in Britain from 1996 onward, but in 2012, the Conservative-Liberal Democrat coalition introduced the "most punitive sanctions ever proposed by a British government" (Tom Slater, cited in Fletcher and Wright 2018, 333), with greater obligations imposed on claimants and greater payment withdrawals. Low-level "offenses," such as not turning up for an appointment, could trigger week-long, month-long, or even open-ended payment stoppages (Tom Slater, cited in Fletcher and Wright 2018, 333).

23. Saul cites Eckhart Tolle's *The Power of Now* (1999) as particularly influential and follows Advaita Vedanta—a Hindu doctrine which proposes a total unity of self and universe.

24. This interrelation is explored at length in chapter 7.

25. Distributed in both physical and digital form, the advert reads, "We need to impress . . . So dress your best," below a picture of two headless male and female monochrome suits.

26. This is the headquarters of Atos, a French IT company awarded the contract by the Labour government in 2008, to assess the right of claimants to sickness benefits in a "Work Capability Assessment" (WCA). After 2011, as part of the new government's austerity program, every long-term recipient of incapacity benefit and every disabled person on income support was to be reassessed by Atos under new criteria, with the expectation that 23% would be found "fit for work" (Department for Work and Pensions 2010, 12). In a Commons debate in 2013, MPs revealed that 1,300 people had died after being told they were able to work after their WCA (Gentleman 2015). In March 2015 the contract was transferred to US company Maximus.

27. Department for Work and Pensions 2010, 12.

28. Like Anonymous, the Skinhead movements of twentieth-century Britain possessed both far right, far left, and apolitical strands. The first wave of Skinheads emerged in the 1960s as a working-class alternative to both right-wing Conservativism on the one hand, and middle-class hippies on the other, and incorporated clothing styles and music brought with the Windrush generation of postwar Caribbean migrants, particularly from Jamaica. During the 1970s economic slump, a second wave of Skinheads emerged, again out of

existing forms of working-class solidarity but this time more expressly political, incorporating both far left and far right factions (Peacock 2007). By the early 1980s, the period Gregor is recalling, Skinheads were largely associated with neo-Nazism and racially motivated violence. The term Skinhead derives from their hairstyle which was often close-cropped or shaved.

29. "Workfare" (an amalgam of "work" and "welfare") existed before the Conservative-Liberal Democrat coalition came to power but was diversified and extended into a number of different "work experience" schemes in 2011 (Ball 2012). One of the most controversial of these was "Mandatory Work Activity." Those who had been unemployed for three months or more would be obliged to fulfil a six-to-eight week placement for thirty hours a week in a business, charity, or other organization, in order to keep their benefit payments. Because of the unpaid and coerced nature of the work, it was compared to slavery or indentured servitude by citizens' rights activists (see www.boycottworkfare.co.uk). The scheme was scrapped in 2015.

30. A zero-hours contract is one in which the employer does not guarantee fixed working hours, hiring staff if and when they are needed. The result is that the employee has no guaranteed income and often no right to sick pay, holiday pay, and other employee benefits.

31. DWP is the acronym for the UK Government's Department of Work and Pensions.

32. Wallace 1956.

33. Wallace 1956, 269.

34. Wallace 1956, 269.

35. Wallace 1956, 264.

36. Male puberty rituals among the Wogeo of Papua New Guinea involve a practice of making the initiands' tongues bleed by rubbing them with coarse leaves. Once the bleeding has stopped, their tongues are then wiped with soft leaves and they are presented with a drink made of cordyline leaves and coconut juice to promote healing (La Fontaine 1985). In the female puberty rite of the Pokot of Western Kenya, newly circumcised women are subject to certain restrictions including washing in the irrigation channels (La Fontaine 1985, 168). Initiands into the Poro male secret society of West Africa are lacerated on the back near the start of the rite, with the time spent in the bush 'partly to give the cuts . . . time to heal' (La Fontaine 1985, 98).

37. La Fontaine 1985.

38. Granovetter 1973.

39. Prohibitions against sleeping, eating, and drinking for the first three or four days are documented among the Yamana of Tierra Del Fuego, and across the Pacific Northwest (Eliade 1995). In Australia, Yuri-ulu initiands are constantly shaken so they cannot fall asleep, while among the Narriniyeri—as with the Orokaiva—they are taken to the bush where they are prohibited from eating or sleeping (Eliade 1995). Wogeo men of Papua New Guinea refrain from sexual intercourse around their contact with the sacred flutes, sometimes for up to nine months (La Fontaine 1985). Vomiting may be induced by a variety of natural emetics, such as the use of salt water by the Native American Kwakiutl (Henderson 1967), and the widespread use of tobacco-based emetics across Native North America (Winter 2000).

2. DREAM

1. As Kathryn Schulz (2015) observes, if you search for "YouTube" and "-hole" you generate over three million results.

2. Gibson first used the term cyberspace in a novella published in 1982, but it was after it appeared in his novel *Neuromancer* (1995) that the term entered the popular lexicon.

3. Van Gennep 1977, 21. He also calls these phenomena "liminal rites" (1977, 21) and "transition rites" (1977, 11) interchangeably.

4. Dave 2014; Chua and Grinberg 2021; Chouliaraki 2006; Cottee 2022; Oliver 2001; Peters 2001; Richardson 2020; Shapin and Schaffer 1985.

5. Peters 2001, 709.

6. Alongside death imagery, uterine symbolism is widely documented during the liminal period of initiation, indicating a time before the ritual subject's rebirth.

7. This is not only the "red pilling" of QAnon, but also the audience cult developed around Andrew Tate in which all political opposition is represented as part of "The Matrix" (Shea 2023).

8. The GCSE (General Certificate of Secondary Education) is a series of school exams normally taken at the age of 16.

9. See chapter 3 for a longer discussion of this doctrine.

10. All except for one key participant discovered Anonymous online. In this outlier, the woman was attending a protest in support of Wikileaks when she fell into conversation with an Anon. He took her email address, said he would add her as a friend on Facebook, and handed her a Guy Fawkes mask with the greeting, "Welcome to Anonymous."

11. Between 2011 and 2013 this march was called #OpVendetta.

12. This is the same video mentioned in the opening pages.

13. Knappenberger 2012.

14. Cf. Anderson 1983 for a thorough exploration of the link between media and religious or societal consciousness.

15. A particularly rich account, retold by Taussig (1999) and summarized at the start of chapter 4, comes from Martin Gusinde, a German-Austrian anthropologist and priest, who was initiated by the Selk'nam of Tierra del Fuego in 1923. This is unusual. Many ethnographic descriptions take the form of Peter Mark's (1992) account of the Jola, where crucial details on the learning process have been omitted because access has not been granted.

16. Since the period of fieldwork, new digital tools have been developed by anthropologists to document the smartphone activity of consenting collaborators, see for example EthnoAlly (Favero and Theunissen 2018).

17. A significant amount has been written about the social and cognitive affordances of YouTube, which forms a complement to this argument: see particularly Munn 2019, 2023; Lange 2016, 2019; and Costa et al. 2022.

18. For a description of YouTube as an archive see Gehl 2009. For several discussions and critiques of this idea see Snickars and Vonderau 2009.

19. In 2008, students at the Massachusetts Institute of Technology built a function named "YouTomb" to track videos removed by YouTube, although this is no longer operating. Deleted YouTube videos can also be searched by using the Wayback Machine supported by archive.org. Luke McKernan (2016) estimates that a quarter of all videos may now be lost.

20. Kavoori 2011.

21. Fisher 2022; Odell 2020; Williams 2018.

22. Srnicek 2016.

23. Benjamin 1970.

24. Benjamin 1970.

25. The "dreamworld of mass culture" is developed by Susan Buck-Morss (2000) in her exegesis of Benjamin's *Arcades Project*.

26. Some of the most extreme effects of audiovisual instruction are documented in Cottee 2022; Martino 2018; and Vacca 2019.

27. Besides material references to sight through the use of masks and unmasking, the knowledge gained can be represented in discursively visual terms. In the Bakhimba secret society of the African Congo, initiands take a solemn oath to secrecy about what takes place at this crucial moment: "All that I see here I will tell to no-one" (cited in Eliade 1995, 75).

28. Ritual practices surrounding the face and eyes are discussed at greater length in chapter 4.

29. Mark 1992.

30. Mark 1992, 94, 104. Some Ejumba masks also contain mirrors on either side of these cylinders which are said to reveal the presence of witches, extending this visual metaphor.

31. Turner 1967, 102.

32. Turner 1967, 103.

33. Turner 1967, 102.

34. McGrain 2011.

35. Turner 1967, 102.

36. Shapin and Schaffer 1985.

37. On Robert Boyle's use of legal analogy see Shapin and Schaffer 1985, 56–57.

38. Shapin and Schaffer 1985, 60.

39. Oliver 2001.

40. Dave 2014.

41. Dave 2014, 442.

42. Richardson 2020.

43. The following description is drawn from the Facebook page of the Cambridge UK chapter I observed in 2018. "Anonymous Voices for the Voiceless . . . is an animal rights organization that specializes in educating the public on animal exploitation and fostering highly effective activist communities worldwide . . . We do not promote or condone anything less than veganism . . .Through the use of standard-practice footage, we expose the public to the TRUTH [sic] behind industries of animal exploitation. Combining this with a value-based sales approach and informational resources, we fully equip the public with everything they need to be vegan and become an active voice for animals." (https://www.facebook.com/groups/285039908591876).

44. While I do not venture a psychoanalytic explanation here, Sigmund Freud's The Interpretation of Dreams (1954) and Eric Fromm's The Forgotten Language (1957) offer concepts of dream experience that resemble the way in which Anons narrate their virtual encounters.

45. The Kunapipi Australian secret cult involves spending a "dreaming period" inside the belly of a snake (R. M. Berndt, cited in Eliade 1995, 48), after which the initiand is smeared with blood and ochre like a new-born, and his spirit "comes out new" (R. M. Berndt, cited in Eliade 1995, 50). The Fox society of Native North America also equate transformation with waking. On the final evening of a nine-year initiation, the initiands go to sleep on the floor of the dance house and "awaken as men" (Van Gennep 1977, 183).

46. During his wanderings the Siberian shaman is said to meet humanoid and animal spirits who reveal secret doctrines and healing techniques (Eliade 1995, 90). In Australia, the Ural-Altaic shaman withdraws to the woods where he likewise encounters a series of benevolent and malevolent spirits offering teaching (Van Gennep 1977, 108).

3. SOCIETY

1. Anonymous Immagical (2014).

2. GCHQ stands for Government Communications Headquarters. NSA stands for National Security Agency, its US counterpart. Five Eyes is agreement on intelligence-sharing between Australia, Canada, New Zealand, the United Kingdom, and the United States.

3. See Barkun 2003; Stewart and Harding 1999; Robertson 2016; West and Sanders 2003; Birchall and Knight 2022.

4. West and Sanders 2003, 6.

5. West and Sanders 2003, 6.

6. Alberoni 1984, 71.

7. Robertson 2016, 201.

8. Patterson 1982.

9. Stark and Bainbridge 1985, 6.

10. Robertson 2016, 26.

11. The protest is also promoted through the main events website, www.ukanonymousevents.com, which is no longer live.

12. I archived this public page as screenshots into a text document. The full data set from which the following analysis is drawn extends across ninety-seven pages of screenshots.

13. The normal statement runs as follows: "We are Anonymous / We are Legion / We do not forgive, We do not forget / Expect us."

14. This video is no longer available. Its original address was http://www.youtube.com/watch?v=-YAbqh2T9OY.

15. Ball 2013.

16. This is a concatenation of several British and American security services, namely: GCHQ, the Federal Bureau of Investigation (FBI), Military Intelligence, Section 5 (MI5), and the Central Intelligence Agency (CIA).

17. Bilderbergers are the anonymous participants in an annual conference initiated in 1954, to strengthen free market links between Europe and North America, which includes people of prominence in various professions. The Bilderberg Meeting is named after the first location in which it was held, the Hotel de Bilderberg in The Netherlands. The Rothschilds is a reference to a banking dynasty dating back to the eighteenth century, which today remains involved in a number of economic concerns, most notably Rothschild & Co. Investment Bank.

18. Alberoni 1984.

19. Alberoni 1984, 72.

20. Alberoni 1984, 29.

21. Robertson 2016, 209.

22. This form of reification is not new. Indeed, an old historical precedent is furnished by the English yeoman parliamentary reformer William Cobbett. In the early nineteenth century Cobbett extemporized against what he called "the Thing": a mixture of politicians, churchmen, aristocrats, and businessmen whom he believed were corrupting British society from within (Osborne 1966, 48).

23. Kersey 2010.

24. See Kent 2015 and Kivanç 2016. Robert Arthur Menard, who remains a guru in the movement, reportedly coined the term "Freeman on the Land" in 2004 (Library of Congress n.d.).

25. Kent 2015.

26. It should be noted that the British Freeman movement has always been ideologically nonviolent, which may partly stem from the pervasive influence of John Harris (discussed later in the text), who stresses the importance of nonviolence in his talk.

27. The activist, referred to as "commonly known as Dom" (a characteristic Freeman construction which disavows title and surname), uses the article as a chance to promote Freemen ideas, particularly the association between birth certificates and enslavement (Playford, Howard, and commonly known as Dom 2011). A video on YouTube shows Dom giving a short presentation to his fellow activists at Occupy London with a poster of a Guy Fawkes mask behind him (FTMRecords 2011).

28. Scobie 2011.

29. This is a reference to one of the most arcane parts of Freeman thinking, drawing on the Cestui Que Vie Act of 1666. This historical act, which sought to reclaim the tenancies of those who had left the country or were lost at sea, is taken as proof that all British citizens have been rendered legally dead so that their land can be seized.

30. Patterson 1982, 5.

31. Patterson 1982, 5.

32. See Stewart and Harding 1999 for an extensive review.

33. Barkun 2003, 40.

34. West and Sanders 2003.

35. Pat Robertson, *The New World Order* (1991). For artifacts of pop culture see Stewart and Harding 1999.

36. Stewart and Harding 1999, 295.

37. Birchall 2006.

38. Premillennialists believe that the millennium will not begin until after Christ's return, and therefore share a catastrophic vision of the great battle between good and evil that precedes this. This is in contrast to postmillennialists, who believe that Christ will come after the millennium has ended, and therefore that the world will be gradually perfected (Barkun 2003). Anonymous contains secular discourses of both millennial concepts—for further discussion see chapter 7.

39. Satan's coming dominion was understood to be a restoration of the Roman Empire. This element is also manifest in Freeman suspicion of Roman law and the use of capital letters on birth certificates to signify slavery, as when Pete says, "back to Roman times."

40. Barkun 2003.

41. Concerns about 5G broadband may be considered the most recent expression of this. In fact Pete himself, fifteen months before these acquired a popular character during the Coronavirus pandemic, expressed concerns about the dangers of the roll-out. See Birchall and Knight 2022 for a more in-depth discussion.

42. The Knights Templar was a Catholic military order founded in the twelfth century. Skull and Bones is an undergraduate secret society at Yale University.

43. Barkun (2003) focuses on the writings of two English women in the 1920s and 30s, Nesta Webster and Lady Queenborough.

44. First among these was the prominent American Nazi sympathizer Gerald Winrod, followed more recently by the Reverend Jerry Falwell in 1999 (Barkun 2003).

45. Sullivan 2009.

46. Sullivan 2009.

47. See Dice 2005 and Jones 2007. The Georgia Guidestones were partially destroyed by an act of vandalism in 2022, and were not rebuilt.

48. In the early stages of the pandemic, Patrick expresses wonder to me that everything he had predicted five years beforehand is becoming reality. The film *Plandemic* (cf. Birchall and Knight 2022) is also widely circulated.

49. Stark and Bainbridge (1985) argue this is an essential quality of cults.

50. Coleman 2014, 31. While legally many of these discourses would be defined as hate speech, here I follow Udupa's (2019) conception of "extreme speech," which emphasizes the social productivity of speech acts, particularly insider/outsider distinctions.

51. Grant 2007.

52. This episode and its implications are further analyzed in Peacock (2025).

53. Nippert-Eng 2005.

54. Nippert-Eng 1996.

55. #AnonFam is a searchable hashtag with many thousands of tweets, an abbreviation of Anonymous Family.

4. MASK

1. Cited by Martin Gusinde in Taussig 1999, 129. This description is drawn from Taussig's retelling of Gusinde's ethnography with the Selk'nam of Tierra del Fuego in 1923.

2. See Frazer 1998; Boas 1927; Fagg 1967; Lévi-Strauss 1983; Goody 2000.

3. Call 2008, 154.

4. Coleman 2012.

5. See Gell 1975 on the Umeda fertility ritual, Jonaitis 2006 on Tlingit shamanic healing practices, Errington 1974 on Karavar political ritual. See Argenti 1997, Koloss 1988, Raposo 2005, and Steiner 1992 on carnivals and masquerades.

6. See Crumrine 1983, Emigh 1996, Tonkin 1979.

7. See Gell 1998 for a theory of artistic action.

8. Frois 2009, 9.

9. As part of ritual procedures of symbolic death, initiands are sometimes given initiation names. See Alford 1988, 87 on the presence of initiation names among the Hopi, Bororo, Tiv, and Cuna.

10. The Selk'nam case above is a representative example. Likewise Eliade (1995, 33) describes the "terror of the women and the uninitiated" at the sight of masked Elema men in New Guinea; Bloch (1992, 9) the "terrifying" emergence from the forest of Orokaiva wearing masks decorated with bird feathers and pigs tusks in Papua New Guinea; Barth (1975, 51) the "shivering, frightened age-mates" swept into the forest by Baktaman men wearing head-dresses and pigs tusks in New Guinea.

11. Turner 1969, 96. Turner introduces this concept in *The Ritual Process* (1969) and returns to it again in his later work, particularly *Dramas, Fields and Metaphors* (1974).

12. Turner 1969, 97.

13. Turner 1969.

14. In a phrase that could also describe male Anons, he says, "No longer were they grandsons, sons, nephews, but simply anonymous novices" (Turner 1974a, 201).

15. Among the Salish of the Native American Northwest, their Swaihwé ceremonial masks were transferred by inheritance or marriage between a limited number of noble lineages, thought to enrich those who wear them (Lévi-Strauss 1983). The Tal mask, by contrast, could be purchased, and was thus a way to acquire social status (Lévi-Strauss 1983).

16. Among some Jola communities of the Casamance region of Africa, only those who demonstrated the skill of clairvoyance during their initiation were permitted to wear the horned Ejumba mask (Mark 1992). For the Poro, a secret society of Sierra Leone, there were up to ninety-nine different ranks of membership, each of which was symbolized by a particular mask whose power was conferred to the mask's owner (La Fontaine 1985, 94).

17. This has been documented among the Jola (Mark 1992, 36–38), the Orokaiva of Papua New Guinea (Bloch 1992, 9). See also Jamieson 2007, 264–66 on the masked puberty ritual among the Miskitu of Nicaragua for something very similar.

18. Whether it is politically as powerful is a different question. Because of the radical fragmentation of the experiences that lead people into Anonymous, there are fewer structural commonalities that connect them. For instance, those who possess a common

labor relationship may share similar kinds of embodied experience, while those who share ethnic, gender, or sexual identities might inhabit a common relation to a particular legal or institutional apparatus.

19. Turner 1974a. La Fontaine (1985, 142) says of her own ethnography among the Gisu of Uganda, "Those who were initiated the same year are age-mates (*ba-magoji*). They are expected to have close ties to one another, especially if they have shared the same convalescent hut." Mark (1992, 57) compares the sense of community forged during *bukut* initiation rites in the Casamance region of Africa to "veterans of a military campaign."

20. Napier 1986. When Anons do pull their masks down over their faces, it is for a variety of reasons. It may be to shield their identities in photographs or videos, or when directly confronting an authority such as at the protest outside GCHQ. In the large majority of cases, this is a temporary measure and will usually soon be followed by forehead-masking.

21. Besides Manchester, marches for the homeless took place on the same day in London, Hull, Brighton, and Norwich, with people wearing Guy Fawkes masks present at several of these. This march was promoted on YouTube, Twitter, and Facebook by a number of people who self-identify as Anon, some of whom I interviewed.

22. As the commercial distributor of *V for Vendetta*, Warner Bros. and its parent company Time Warner own the rights to the image of the Guy Fawkes mask and are paid a licensing fee every time a mask is sold (Bilton 2011). This jaundiced version is a Warner Bros. production. More popular is a simpler, whiter version which has been produced without this license.

23. Before this event took place, Manchester City Council approved £59 million in budget cuts (Howson 2014).

24. Bedroom Tax became the colloquial term for the Welfare Reform Act passed in 2012. It entailed a 14% reduction in Housing Benefit if tenants were deemed to possess more than their calculated allowance of rooms. The tax was purportedly introduced to ensure council tenants were housed in accommodation commensurate with their needs, but in practice it had a disproportionate effect on disabled people, who made up two-thirds of the tenants affected (Butler 2014).

25. Manchester Town Hall was built in 1877 and intended to reflect the wealth and status the city had acquired as an industrial center. In contrast to contemporary doxa of budgetary constraint, it was to be built "at any cost" required (cited in Hartwell 2002, 71).

26. For a journalistic account of this event, see Williams 2015.

27. See chapter 8 for more detail on the mechanics of a sleep out.

28. The square outside Manchester Town Hall is called Albert Square, named after Prince Albert, the consort of Queen Victoria, who is materialized here in larger than life-size marble.

29. This camp persisted in Albert Square after the march, subsequently moving to several other sites before being finally evicted in September of that year (BBC News 2015b; Quays News 2015). After just a month the camp had already cost the Greater Manchester Police an extra £88,000 in policing, security, and legal costs (BBC News 2015a).

30. Council Tax is a compulsory tax that falls on all those over the age of eighteen who rent or own a home (with some exceptions). The license fee is also compulsory for anyone who uses a broadcast television (which also has a number of exceptions).

31. Mauss 1985.

32. Carl Jung (1953, 190) describes the social persona as "a kind of mask, designed on the one hand to make a definite impression on others, and on the other *to conceal the true nature of the individual* (emphasis added). In *Black Skin, White Masks*, Franz Fanon (1970) draws on Alfred Adler's notion of the inferiority complex, to argue that Black Antilleans seek to take on the identity of Whiteness in denial of their own true identity as Black. Rather like Simon and Mr. Roberts, these masks become symbolic vehicles of oppression.

33. On the naked face as a metaphor for truth see Taussig 1999, 223–48.

34. The fuller versions of each quote can be found at the beginning of this chapter and chapter 3.

35. Taussig 1999.

36. For instance, Hannah Arendt (2006, 96) describes a scene from the French Revolution in which revolutionaries express their wish to tear off the "mask of hypocrisy" and expose the "honest face of the people."

37. This form of defacement is likewise documented among the Native American Hopi (La Fontaine 1985, 88–90). Although we do not have a record of the precise moment of defacement, we could presume its occurrence in the rites of the Elema secret society of New Guinea (Eliade 1995, 33); and the Orokaiva (Bloch 1992, 8–14) and Baktaman of Papua New Guinea (Barth 1975, 51), as they all involve the terrorization of initiands by masked beings before the former earn the right to wear the masks themselves.

38. Freemasons are stripped and blindfolded and go through various physical ordeals in this state (incarceration, strenuous exercise, burning, blood loss) before the blindfold is loosened and their passage into Masonry complete (La Fontaine 1985, 49–51); Wogeo of Papua New Guinea are blindfolded and slithers of bone pushed through the lobe and top of each ear (La Fontaine 1985, 132); Australian Yuin have their eyes covered before one of their incisors is knocked out with a chisel (Eliade 1995, 11–12); Australian Karadjeri are blindfolded before they are circumcised (Eliade 1995, 22). Malawi boys are blindfolded before being taken to the initiation site (Joseph and Bonongwe 2005, 16).

39. On initiation into the Chinese Triads a cockerel is beheaded and some of its blood mixed with the blood of the initiand, marking their entry to the society as blood brothers (La Fontaine 1985, 46–7). The Kenyan Mau Mau consume goat's eyes while taking their oath of loyalty (La Fontaine 1985, 77).

40. Pre-Columbian burial masks have been excavated across the Americas which had been placed not over the faces of the deceased, but in the ground alongside them (see Oosten 1992 on the Alaskan Ipintak and Markman and Markman 1989, 88–92 on Oaxaca and Mexico). It may be speculated that it is only deities or other immortal creatures, such as the pharaohs of Egypt, who are buried inside whole body masks (in the form of sarcophagi), as the paradox presented by the biological death of human beings does not obtain in the same way.

41. The masks are nine thousand years old and thought to be modeled on human skulls (BBC News 2014b).

42. Turner 1974b.

43. Handelman 2021, 66. See also Handelman 1990.

44. Taussig 1999, 249.

45. Cases of covering and adornment to mark the end of an initiation rite are ubiquitous across the literature. Besides the examples of mask-wearing cited above, the following are a handful of representative illustrations. Just before the completion of the fourth stage of initiation, New Guinea Baktaman novices have their hair elaborately tied with string and barkcloth, their bodies stained red, and white bird-feathers attached to the tops of their heads (Barth 1975, 73–4); similarly the Barasana of Colombia are also painted red, with string beads tied around their necks, garters placed around their calves, and feather crowns placed on the tops of their heads (Hugh-Jones 1979, 96). Sometimes these coverings cloak the entire body or even constitute the body itself. On entry to a German men's secret society, members ritually don a wolf-skin in imitation of the wolverine hero in an ancient Germanic saga (Eliade 1995, 77); while a Siberian shaman is said to be demonically dismembered during the transitional phase, before spirits cover his bones with new flesh, and in some cases infuse him with new blood (Eliade 1995, 90).

46. See chapters 5 and 8 for further examples.

47. Turner 1986.

48. John Harris's call for the mass immolation of birth certificates (chapter 3) becomes relevant here.

49. With the notable exception of Taussig 1999, the scholarship on initiation masking is saturated with the language of fakery. La Fontaine (1985, 88) says, "the beings who appear in rituals are not spirits but men who impersonate them"; Eliade (1995, 11) that a Yuin initiation in Australia involved "masked and disguised men."

50. See Markman and Markman 1989, on how ritual masks became important sites of conflict in the battle for the real in South America.

51. Cited in McGovern 2013, 170. This particular program takes place in the African state of Guinea.

52. Benson 2006.

5. KNOWLEDGE

1. This description is abridged from Audrey Richards's (1988) classic account of the *Chisungu*, a female puberty rite that took place in what is now Zambia in June 1931. At the time, Richards observed that the rite was already dying out under the impact of the colonial presence, and one of the reasons this ceremony took place at the time and place that it did was because she was there and able to acquisition some of the necessary produce. Formerly, she records, the ceremony could have lasted up to a year.

2. Cited in Richards 1988, 125.

3. Cited in Richards 1988, 121.

4. Karl Popper (1959) argued that what defines scientific knowledge is its capacity to be falsified, shown to be untrue by way of an empirical test. By contrast, other forms of knowledge may be "unfalsifiable," a word sometimes applied to the knowledge of vernacular theorists (Bowen 2018).

5. See Van Badham 2021, 71 on "epistemic provocation." See notes below for further references.

6. See Birchall 2006 on the 1990s popularization of alternative knowledge. See Birchall and Knight 2022 for a more recent discussion.

7. Procházka and Blommaert 2021, 24.

8. Munn 2023, 79.

9. Munn 2023, 78.

10. Munn 2019, 2023.

11. Eliade 1995, x; La Fontaine 1985, 15.

12. For a chronology of the Assange legal case, see Stack, Cumming-Bruce, and Kruhly 2019.

13. The rejection of organization often requires spontaneous adjustments. The advertised intention is to gather in the main plateau of Trafalgar Square, just as the Million Mask March did. However, on arrival it becomes clear that this is already being filled by a large Russian folk festival. The event then moves to this traffic island on the south side of the square instead.

14. An entry in the Urban Dictionary defines twerking as "The rhythmic gyrating of the lower fleshy extremities in a lascivious manner with the intent to elicit sexual arousal or laughter in one's intended audience" (The Prince of Diamonds 2012).

15. The "Chemtrails" theory maintains that the vapor trails left by some flying aircraft are hazardous chemicals, sprayed across the earth's atmosphere for nefarious purposes. For a longer discussion of this theory in the context of Greece, see Bakalaki 2016.

16. It should be remarked that this robust leadership style is unorthodox for Anonymous and is consequently the subject of leveling humor.

17. For a critical engagement with such knowledge practices see Gray, Bounegru, and Venturini 2020.

18. Birchall and Knight (2022, 128) call this "epistemological flatness."

19. UK's Office for Standards in Education, Children's Services and Skills reports that in 2015 approximately 45% (258,528 of 571,334) of all schoolchildren in Britain continued to post-16 education.

20. See Uscinski and Parent 2014.

21. See **table 1** for a full list of documentaries cited.

22. Simon's encounter with the Licensing Agent, related in chapter 4, exemplifies some of the social interactions that can arise from this boycott.

23. See **table 2** for further details of this event.

24. The contradictory attitudes of Anons toward the BBC appear again later in the portraits of Garvey and Betty, both of whom use the broadcaster to advance their political messages.

25. Jola learn how to make their horned mask costumes for the Bukut initiation ceremony (Mark 1992, 106). Orokaiva learn how to play flutes and bull-roarers (Bloch 1992, 9).

26. Jola also learn certain oral traditions, songs, and particular signals through which to recognize other initiates (Mark 1992, 49). Orokaiva are taught to play sacred flutes and perform spirit dances (Bloch 1992). Initiands into the Duk Duk Secret Society of Melanesia are taught a sacred dance (Van Gennep 1977, 82). Initiation into Muslim brotherhoods in Cairo entails the learning of sacred formulas (1977, 97).

27. Turner 1967, 105.

28. Richards 1988, 125. The Bemba live in present-day Uganda. Similarly, La Fontaine (1985, 169) reports that Pokot girls are harangued by their mothers during their puberty rite, a practice explicitly designed to "teach" (169) them as preparation for marriage.

29. See examples of oath-taking in indigenous Australia (Eliade 1995, 10); among the Mau Mau secret society of Kenya (La Fontaine 1985, 75); among the Nyoro of Western Uganda (La Fontaine 1985, 64); and among a whole variety of European and extra-European secret societies (Simmel 1906).

30. Such an oath can take the following form, documented among the Bakimba of Nigeria: "All that I see here I will tell to no-one, neither woman, man, non-initiate, nor white man; otherwise make me swell up, kill me" (cited in Eliade 1995, 75).

31. Simmel 1906, 482. This is a translation of Simmel's German term *Heraussonderung*.

32. Cited in Eliade 1995, 37. In a subsequent discussion of the significance of the mysteries in primitive Christianity, Eliade (1995, 120) observes that the phrase "this is known to the initiates" runs through the Greek sermons on the subject.

33. Goffman 1962, 4.

34. Goffman 1962, 14. Through his concept of mortification Goffman retains an emphasis on the death-like aspect of passage rites.

35. La Fontaine 1985.

36. Cited in La Fontaine 1985, 159. La Fontaine carried out fieldwork among the Gisu of Uganda.

37. Eliade (1955) calls this narrative pattern "the myth of the eternal return." In contrast to the linear and progressive sequences of historical time, these ways of configuring temporality are fundamentally cyclical and regenerative, seeking to recover a golden era that has been lost. See chapter 6 for a longer exploration of this dichotomy.

38. Respectively, these sites are: the center of Indianapolis, US; the area around the White House in Washington, US; Buckingham Palace in London, UK; Hampton Court Palace, UK; Karlsruhe Palace in Germany; the Palace of Versailles, France; Vatican City;

Rome; the center of Cairo, Egypt; the center of Baghdad, Iraq. This video is no longer publicly available. The original URL is https://www.youtube.com/watch?v=Sty9t9A3Fok

39. Played by Michael Douglas, Gordon Gekko is the central character of the film and most famously associated with the phrase "Greed is Good" (Stone 1987).

40. To go into service means to become a domestic servant in a household. In the early twentieth century domestic service was a major employment sector.

41. Sparrow 2013.

42. The subheading, "Knowledge is power, and the 99% are waking up," is sourced from a visual meme created by Tank.

43. For an in-depth discussion of this idea and its social effects, see chapter 7.

44. In this discussion of value and its embodiment, I am pursuing some of the theoretical claims made by Louis Dumont. Dumont (1986, 25) introduced the idea of what he called "paramount value," a value so central to a given community that it inflects every aspect of it in some way. At a sociological level, Dumont argues that it is those persons closest to representing the paramount value who enjoy the highest status within it. Although Anonymous is ideologically acephalous, a tacit hierarchy corresponding to the value of exposure is presented here.

45. The existence of the Official Secrets Act in Britain—in which certain persons are required by law not to disclose certain sensitive information—constitutes those persons as members of a de facto secret society. MI5 employees are compelled to sign a statement agreeing to the provisions of the Act, which functions as the bureaucratic equivalent of the vows and oaths documented by Simmel (1906).

46. On the doctrine of revelation see Dulles 1992.

47. The 1978 Chicago Statement on Inerrancy reads, "If this total divine inerrancy is in any way limited or disregarded . . . such lapses bring serious loss to both the individual and the Church."

48. Nagel 1986.

49. Steiner 1991, 68.

50. See Steiner 1991, 126 for a more detailed account of this historical shift that he calls "the great spiritual change."

51. Müller 2020.

6. SYMBOL

1. McTeigue 2005.

2. For further illustrations see Burridge 1969; Lindstrom 1993; Steinbauer 1979; and Worsley 1957.

3. This description is drawn from Burridge 1969, 15–17.

4. This has also been called "ingratiation" (Tourish and Wohlforth 2000, 29), or in the Unification Church, "heavenly deception" (Barker 1984, 176). For a further description and discussion of love-bombing see chapter 7.

5. Turner 1967, 20. The Latin name for this species is *Diplorrhyncus condylocarpon*.

6. For ethnographic accounts of the Houses of Parliament see Crewe 2015 and Cockerell 2016.

7. At the time of research, one of the most significant of these is Occupy Democracy, see Graeber 2013.

8. For an earlier discussion of defacement relative to masking practices see chapter 4.

9. Taussig 1999, 1.

10. Thompson 1967.

11. Ogle 2015.

12. McTeigue 2005.

13. Discourses of the present also appear in chapter 3: Outside GCHQ.

14. Benjamin 1970.

15. Turner 1969, 96. For a fuller citation see the epigraph at the beginning of the Introduction.

16. See Irvine 2020 for a geological history of this region.

17. This industrial history of Peterborough is adapted from the following secondary sources: Bendixson 1988; Brandon and Knight 2001; Labrum 1994; and Tebbs 1979.

18. Unemployment was 15.3% in 1983 (Bendixson 1988).

19. The largest of the latter is Gateway Peterborough, a 240-acre site which first opened in 2004.

20. The Newsroom 2017.

21. This enterprise was known as the Gregorian Mission, having been initiated by Pope Gregory.

22. The Anglo-Saxon chronicle was produced here in the twelfth century and contains what is thought to be the first-ever written record of the pronoun "she."

23. Tebbs 1979, 70.

24. There are also OpRealLove events taking place on the same day in Manchester and Liverpool, administered by different groups of Anons.

25. Marx 1995, 42–50.

26. On refacement see chapter 4.

27. Many years after this event, Garvey went on to win his case with a full exoneration, in what was described as one of the greatest miscarriages of justice in British history.

28. For a previous discussion of Pierre Bourdieu's concept of the "Left" and "Right hand" of the state, see chapter 1.

29. NHS revenue spending on private providers rose to 10.7% in 2015/16. Source: full-fact.org.

30. Benjamin 1970.

31. Eliade 1955.

32. Eliade 1955, 158.

33. Eliade 1955, 35.

34. Cited in Cressy 1992, 71.

35. Cressy 1992.

7. GROWTH

1. McTeigue 2005.

2. Barkun 2003, 56.

3. See particularly Barker 1984; Dawson 1998; and Robbins 1988.

4. There is substantial dispute over whether Pelagius actually espoused many of the ideas attributed to him, comprehensively addressed by Bonner (2018). For the purposes of this discussion the question is not significant, as even if Pelagius were merely a doorpost on which subsequent generations of theologians hung their ideas, it still demonstrates the growing appeal of these ideas in the European canon.

5. Condorcet 1955.

6. Of these, the most prominent is www.collective-evolution.com

7. Lane 1996.

8. Hubbard 2015.

9. Marilyn Ferguson, in her book *Aquarian Conspiracy* (1982), interviewed 185 nominal leaders of the New Age movement. Among these, de Chardin is cited as the most important intellectual influence overall.

10. Cited in Delfgaauw 1969, 13.

11. Cited in Roberts 2000, 143.

12. Cited in Roberts 2000, 142 and Lane 1996, 123 respectively.

13. Lane 1996.

14. Cf. Srnicek 2016; Zuboff 2019.

15. La Fontaine 1985, 16. La Fontaine's emphasis on the public character of initiation contrasts sharply with Eliade's (1995, 103) insistence that initiation is "first and foremost a secret rite."

16. Cowell reportedly donated $150,000 to Friends of the Israeli Defense Forces at a gala in October 2013, becoming a subject of public controversy on Twitter in July 2014.

17. Alberoni (1984, 132) describes similar familial separations during the youth movements of 1968. "Parents found themselves being rejected in toto for what they were, for what they had been, and for what they had or had not done—rejected whether they reacted harshly or with affection. Their son or daughter was, in fact, asking them to be different without indicating any kind of means-end schema whatsoever."

18. This description is taken from Barth 1975, 51, the opening salvo of the Baktaman male puberty rite in New Guinea. Similarly violent separations have been documented elsewhere. Among the Murring of southern Australia, mothers sit down on the ground covered in blankets with their sons in front of them, when they are suddenly approached by the adult males, who seize the boys before running off together. See also the Australian Wiradjuri and the Yuin for similar examples (all cited in Eliade 1995, 4–8). James Frazer describes entry to the Kaikan men's association on the Indonesian island of Ceram, when the boys are taken blindfolded into a forest hut. While their mothers and relatives look on from outside, a dull chopping sound is heard together with a loud cry, and a bloodied sword thrust through the roof. At the sight of the sword the mothers weep passionately, saying the devil has murdered their son (cited in Bettelheim 1955, 215). The boys are sworn to secrecy about what happens inside the hut.

19. For example, during the Bora ceremonies in New South Wales, after several days of performance, the men, women, and children all form a large circle. The boys are then commanded by their mothers to enter the circle as a sign that they relinquish their authority over them (cf. Webster 1932, 21). At one point in the Kurnai male puberty ritual, the mothers sit behind the boys, and the men walk between the two groups in single file, thus separating them (Eliade 1995, 7). During the rites of the Yaunde of Cameroon, boys fasten banana leaves to their legs which symbolize their enduring femininity, and are then dramatically torn off by female members of the community to break this symbolic link (Webster 1932, 23).

20. This is a reference to the mothers of Yaroinga boys in Australia, whom on seeing their sons emerge from their seclusion flamboyantly decorated with waist-belts and head-dresses, immediately begin to cry and to smear themselves with grease and ashes in mourning (Webster 1932, 21).

21. Arunta boys are told while being painted that they are no longer allowed to play with the women and girls and that they must live in the men's camp (Webster 1932, 22). Among the Yuin it is the women who leave the boys and set up a new camp (Eliade 1995, 7–8).

22. In New Caledonia the boys are circumcised at the age of three and remain with their mothers only until they are weaned, at which point she becomes just another senior woman (Webster 1932, 23). According to an old account, Webster (1932, 23) tells us, Hottentot boys remain with their mothers until eighteen after which point they must assiduously avoid her company.

23. According to Michael Wesch (2008), the first film of a Free Hug was a YouTube video compiled in 2006 of one that took place in Sydney, Australia in 2004.

24. Barker 1984, 173-184.

25. Barker 1984, 103.

26. Coleman 2014, 2.

27. Godiva is a documented historical figure but it is likely that this tale is apocryphal, as it was first recorded nearly two centuries later in the Latin chronicle *Flores Historiarum*. The mythic life of the story, however, is pronounced, particularly from the nineteenth century onward.

28. Cousin Itt is a character from the film and television series *The Addams Family*, principally known for being shrouded in several feet of his own hair.

29. Pers. comm. At an emergency protest in Parliament Square against a vote for air strikes in Syria (see **table 2**), a man and a woman make straight for the gathering. As they pass too closely by, the woman says to one of the Anons in a North American accent, "They beheaded our people," intended as a criticism of the protest. The Anon is put out and yells out after them, "I didn't approach you!" pausing briefly before adding, "Wake up!"

30. Granovetter 1973.

31. Granovetter 1973, 1362.

32. This paint was by an artist at the demonstration, whose offer to be daubed she willingly accepted. See **table 2** for more information on each of the protests listed.

33. For more detail on the West Hendon estate controversy see Booth (2015).

34. La Fontaine 1985.

35. Cf. Bettelheim 1955.

8. SACRIFICE

1. "Anonymous makes homeless man cry" (2013).

2. The Anons refer to the gentleman as a "homeless man," and the intervention rests on the premise that he is homeless. It is impossible to corroborate this. All we can attest is that he is begging for money from passersby, hence the crude but empirical nomenclature, beggar.

3. Retrospective efforts by Anons to identify this man suggest he is a Romanian national named Stefan. Born in 1941, they report, he and his family fled to Bulgaria, where in 1947 all of the latter died ("#PirateSec message regarding Steven"; this video was published in 2015 but is no longer available). While I cannot verify the methodology that has yielded this information, the cardboard sign he is holding in the video conveys a few words of English in a style that resembles Romanian script.

4. This premodern history is adapted from Armitage 2015.

5. These included blinding and castration Armitage 2015.

6. In this early incarnation, Robin Hood was considered a "yeoman," i.e., not a serf nor a member of the gentry.

7. There have been numerous studies of the legend of Robin Hood and how it changed over time through print culture, cf. Knight 2003.

8. This labor history is adapted from Armitage 2015; Chapman 1997; and Lowe and Richards 1982.

9. This radical history is adapted from Beckett 1997.

10. Contemporary commentator Charles James Fox described the "uncontrollable spirit of riot" that pervaded the city at election time, cited in Beckett 1997, 284.

11. Beckett 1997.

12. See Bohstedt 1983 on the food riots of the eighteenth century, and the electoral riots of the nineteenth century, all of which centered on the square.

13. Lowe and Richards (1982, 39) state that by 1907, "The lace industry had reached its peak and was employing over 40,000 people in Nottingham alone."

14. "#OpSafeWinter—Nottingham—13th December 2014" (mind open 2014). This event took place on 13 December 2014; see **table 2** for more details.

15. See Coleman 2014 for examples of the language that was cultivated and used in earlier forms of Anonymous.

16. At the time of research the station has 17, 262 followers.

17. For more information on how the story of OpSafeWinter is told see Don 2014.

18. In this document, the first mission is to count the current homeless population in each town, city, and county. The author offers that "You can get the latest figures from your local council by sending them a freedom of information request if you are not sure of the figures in your town" (A Guest 2013).

19. Some of these accounts are SchismUk, MoreYesThanEver (a reference to Scottish independence), and AnonBridgend (addressing the "people of Wales" in its own tweet). All quantitative data are accurate at the time of fieldwork.

20. AjacxUK 2013.

21. The citation is transcribed as written.

22. Turner 1969, 95–96.

23. See chapter 4 for more detail on masked dances.

24. Taussig 2012.

25. The collected writings of Aaron Swartz (2016) provide an emic overview of this concept of self-organization, as well as what could be considered its most substantial and lasting achievement—Wikipedia.

26. Mauss 2016. A particular example of this comes in Andrew Strathern's (1975) study into making "moka," in which Papua New Guinea big-men deploy unequal forms of gift-giving to accrue value to themselves and build their reputation.

27. On this prehistory see Frazer 1998 and Hubert and Mauss 1968; for more recent reflections on sacrifice, see Valeri 1985 and de Heusch 1985. Mayblin and Course (2014, 307) tell us that in the latter part of the twentieth century, anthropology "sacrificed sacrifice," abandoning it almost entirely as a meaningful subject of inquiry.

28. Mayblin and Course 2014.

29. Mayblin and Course 2014, 309.

30. Hubert and Mauss 1968, 25.

31. Titmuss 2002.

32. This proportional increase far outstripped "population changes, increases in the provisions of beds and the number of patients treated under the NHS" (Titmuss 2002, 42).

33. Titmuss 2002, 226–35.

34. Titmuss 2002, 88. On the concept of the free gift and its relationship to anonymity see Laidlaw 2000.

35. The blood sacrifice centers Hubert and Mauss's (1968) analysis, and it has since been critiqued as privileging a Christocentric view of the phenomenon for this reason (cf. Mayblin and Course 2014).

36. Titmuss 2002, 225.

37. Osborne first makes this statement at the Conservative Party conference in 2009, and repeats it throughout his time in office.

38. camdefendeducation 2010. See Introduction for the unequal distribution of austerity policy.

39. Captain SKA 2010.

40. Although Thatcher did not coin this expression, her repeated use of it attached the phrase to her, one which so came to define Thatcherism that it was abbreviated to the

acronym TINA. David Cameron used this expression in a pre-budget speech in March 2013.

41. Franklin 2019. While I am in accord with Franklin's statement, an argument can also be made that all sacrifice is to some degree concerned with reproduction. The phenomenon of agrarian sacrifice in particular, in which first fruits are ritually consumed or destroyed, animals slaughtered, or even the blood of human victims sprinkled over the fields—all to ensure the crops will grow again the following year—continually links sacrifice to this desire for human endurance.

42. Webb 2018.

43. Khomami 2017.

44. Crawford 2014.

45. Piven and Cloward (1977) document the use of eviction blocks during the Great Depression.

46. Mendax 2014.

47. Bourdillon and Fortes 1980.

48. *The Telegraph* 2015; Parveen 2015.

49. The slaughtering of pigs is particularly common: see Barth 1975, 84–90 on the Baktaman of New Guinea; and Gesch 1985, 236–37 on the Mt. Rurum Movement of Papua New Guinea. The Bimin-Kuskusmin of Papua New Guinea sacrifice pandanu fruits and marsupials (Poole 1982, 132). Eliade (1995, 81) documents the hunting of bears and wolves in the ancient German men's secret societies, the *Männerbünde*. Ascending to the highest grades of initiation among the Ilahita Arapesh of Papua New Guinea involved the hunting of human sacrifices before the practice was banned in 1950 (Tuzin 1982, 334–45). Among the Wogenia of Zaire a potsherd is ritually smashed with a stick (Droogers 1980, 296–98).

50. They describe their "overriding concern as . . . the modification of moral persons through a movement towards the other side of sacrifice, whatever, whenever, and wherever that may be" (Mayblin and Course 2014, 317).

51. Turner 1967, 98–99.

52. Orlando Patterson (1982, 45) also theorizes the blending of the latter stages through his concept of "liminal incorporation." He uses this term to explore the ways in which the slave was symbolically killed and then reincorporated into the house of the master, but incorporated as a non-being, that is to say socially dead. The difference between Patterson's approach and the one adopted here is an artifact of a changing political economy. Although liminally incorporated, the slave was still incorporated into a given social system because labor remained fundamental to the slave system. In twenty-first-century Britain, automation, outsourcing, and widespread precaritization create a different set of labor conditions that dilute the drive to incorporate. This has many analogues elsewhere that have been understood through the language of "dispossession" (cf. Harvey 2003).

53. https://www.streetskitchen.org/

CONCLUSION

1. La Fontaine 1985, 72–73.

2. What La Fontaine's discussion does not mention is that British authorities also tried to break these vows of silence in the 1950s, using torture in the prison camps where some Mau Mau were held. This only became publicly known when a tranche of British colonial archives were released (see Elkins 2005). In its own violent way it also reinforces La Fontaine's argument about the potency of the ritual.

3. See Pariser 2012.

4. Handelman 2021, 63.

5. Miller and Horst 2012.

6. Margetts et al. 2016.

7. Burridge 1969.

8. See the photograph taken by Hollie Adams in Rotheram that appeared in the *Financial Times* on 6 August 2024 (Wallis, Gross, and Fisher 2024).

9. See Lazar 2018 for a full elaboration of this proposition.

References

A Guest. 2013. "#OpSafeWinter." Paste Site. Pastebin.com. Accessed 19 November 2024. https://pastebin.com/fTVgVgs1.

AjacxUK, dir. 2013. *Everybody Is Me #OpSafeWinter*. https://www.youtube.com/watch?v=Rufyz8zhZSA.

Alberoni, Francesco. 1984. *Movement and Institution*. Translated by Patricia C. Arden Delmoro. Columbia University Press.

Alford, Richard D. 1988. *Naming and Identity: A Cross-Cultural Study of Personal Naming Practices*. HRAF Press.

Aly, Anne, Stuart Macdonald, Lee Jarvis, and Thomas Chen, eds. 2016. *Violent Extremism Online: New Perspectives on Terrorism and the Internet*. Routledge. https://doi.org/10.4324/9781315692029.

Anderson, Benedict R. O'G. 1983. *Imagined Communities: Reflections on the Origin and Spread of Nationalism*. Verso.

Anon. 2021. *Post-Internet Far Right: Fascism in the Age of the Internet*. Dog Section Press.

Anonymous Immagical, dir. 2014. *Anonymous UK-GCHQ-Protest 30 August 2014*. https://www.youtube.com/watch?v=eCfTOpQQTtw.

Anonymous Makes Homeless Man Cry. (Original). 2013. http://www.youtube.com/watch?v=CyY0Bct6qz8&feature=youtube_gdata_player.

Arendt, Hannah. 2006. *On Revolution*. Penguin Books.

Argenti, Nicolas. 1997. "Masks and Masquerades." *Journal of Material Culture* 2 (3): 361–81. https://doi.org/10.1177/135918359700200305.

Armitage, Jill. 2015. *Nottingham: A History*. Amberley.

Atkinson, Will, Steven Roberts, and Michael Savage, eds. 2012. *Class Inequality in Austerity Britain: Power, Difference and Suffering*. Palgrave Macmillan.

Bakalaki, Alexandra. 2016. "Chemtrails, Crisis, and Loss in an Interconnected World." *Visual Anthropology Review* 32 (1): 12–23. https://doi.org/10.1111/var.12089.

Ball, James. 2012. "Government's Work Experience: What Are the Schemes, and Do They Work?" *The Guardian*, 22 February 2012, sec. Society. http://www.theguardian.com/global/reality-check-with-polly-curtis/2012/feb/22/unemployment-work-programme-welfare.

———. 2013. "Leaked Memos Reveal GCHQ Efforts to Keep Mass Surveillance Secret." *The Guardian*, 25 October 2013, sec. UK news. https://www.theguardian.com/uk-news/2013/oct/25/leaked-memos-gchq-mass-surveillance-secret-snowden.

———. 2023. *The Other Pandemic: How QAnon Contaminated the World*. Bloomsbury Publishing. https://cam.ldls.org.uk/vdc_100187221917.0x000001.

Barker, Eileen. 1984. *The Making of a Moonie: Choice or Brainwashing?* Basil Blackwell.

Barkun, Michael. 2003. *A Culture of Conspiracy: Apocalyptic Visions in Contemporary America*. 1st ed. Comparative Studies in Religion and Society 15. University of California Press.

Barth, Fredrik. 1975. *Ritual and Knowledge among the Baktaman of New Guinea*. Universitetsforlaget, Yale University Press.

BBC News. 2014a. "London 'Guy Fawkes' Protest Sees 10 People Arrested."
    BBC News, 5 November 2014, sec. London. https://www.bbc.com/news/
    uk-england-london-29919083.
——. 2014b. "World's Oldest Masks Go on Display in Jerusalem." *BBC News*, 11
    March 2014, sec. Middle East. https://www.bbc.com/news/world-middle-
    east-26533994.
——. 2015a. "Homeless Evicted from Manchester City Centre Camp." *BBC News*, 18
    September 2015, sec. Manchester. https://www.bbc.com/news/uk-england-
    manchester-34290299.
——. 2015b. "Manchester Homeless Camps Banned from City Centre." *BBC News*,
    30 July 2015, sec. Manchester. https://www.bbc.com/news/uk-england-
    manchester-33717978.
Beckett, John, ed. 1997. *A Centenary History of Nottingham*. Manchester University
    Press.
Bendixson, Terence. 1988. *The Peterborough Effect: Reshaping a City*. Development
    Corporation.
Benjamin, Walter. 1970. *Illuminations*. Edited by Harry Zohn. Cape.
Benson, Susan. 2006. "Injurious Names: Naming, Disavowal, and Recuperation in
    Contexts of Slavery and Emancipation." In *The Anthropology of Names and
    Naming*, edited by Gabriele Vom Bruck and Barbara Bodenhorn, 177–99.
    Cambridge University Press.
Bettelheim, Bruno. 1955. *Symbolic Wounds: Puberty Rites and the Envious Male*.
    Thames and Hudson.
Beyer, Jessica Lucia. 2014. *Expect Us: Online Communities and Political Mobilization*.
    Oxford Studies in Digital Politics. Oxford University Press.
Bilton, Nick. 2011. "Masked Protesters Aid Time Warner's Bottom Line." *The New York
    Times*, 28 August 2011, sec. Technology. https://www.nytimes.com/2011/08/29/
    technology/masked-anonymous-protesters-aid-time-warners-profits.html.
Birchall, Clare. 2006. *Knowledge Goes Pop: From Conspiracy Theory to Gossip*. Culture
    Machine Series. Berg.
Birchall, Clare, and Peter Knight. 2022. *Conspiracy Theories in the Time of COVID-19*.
    1st ed. Conspiracy Theories. Routledge. https://cam.ldls.org.uk/vdc_100165
    357327.0x000001.
Bloch, Maurice. 1992. *Prey into Hunter: The Politics of Religious Experience*. Lewis
    Henry Morgan Lectures. Cambridge University Press.
Blyth, Mark. 2015. *Austerity: The History of a Dangerous Idea*. Oxford University
    Press.
Boas, Franz. 1927. *Primitive Art*. Skrifter8. H. Aschehoug.
Bohstedt, John. 1983. *Riots and Community Politics in England and Wales, 1790–1810*.
    Harvard University Press.
Bonner, Ali. 2018. *The Myth of Pelagianism*. British Academy Monograph. Oxford
    University Press.
Booth, Robert. 2015. "Barnet Council 'Engaged in Social Cleansing' at West Hendon
    Estate." *The Guardian*, 20 January 2015, sec. Society. https://www.theguardian.
    com/society/2015/jan/20/
    barnet-council-social-cleansing-barratt-homes-west-hendon.
Bourdillon, M. F. C., and Meyer Fortes, eds. 1980. *Sacrifice*. Academic Press for the
    Royal Anthropological Institute of Great Britain and Ireland.
Bowen, Jack. 2018. "Unfalsifiability." In *Bad Arguments*, 403–6. John Wiley & Sons.
    https://doi.org/10.1002/9781119165811.ch99.

Brandon, David, and John Knight. 2001. *Peterborough Past: The City and the Soke.* Phillimore.

Britannica. 2024. "tonsure." Accessed 15 February 2025. https://www.britannica.com/topic/tonsure.

Brooks-Pollock, Tom. 2014. "What Is the Million Mask March?" *The Telegraph.* 6 November 2014. https://www.telegraph.co.uk/technology/news/11212615/What-is-the-Million-Mask-March.html.

Buck-Morss, Susan. 2000. *Dreamworld and Catastrophe: The Passing of Mass Utopia in East and West.* MIT Press.

Burridge, Kenelm. 1960. *Mambu: A Melanesian Millennium.* Methuen.

——. 1969. *New Heaven, New Earth: A Study of Millennarian Activities.* Pavilion Series. Social Anthropology. Blackwell.

Büscher, Monika, and John Urry. 2009. "Mobile Methods and the Empirical." *European Journal of Social Theory* 12 (1): 99–116. https://doi.org/10.1177/1368431008099642.

Butler, Patrick. 2014. "Bedroom Tax Has Failed on Every Count." *The Guardian,* 28 March 2014, sec. Society. http://www.theguardian.com/society/2014/mar/28/bedroom-tax-failed-overcrowding-savings.

——. 2015. "Those Food Bank Data: Complicated Yes; Exaggerated, No." *The Guardian,* 4 May 2015, sec. Society. https://www.theguardian.com/society/patrick-butler-cuts-blog/2015/may/04/those-food-bank-data-complicated-yes-exaggerated-no.

Call, Lewis. 2008. "A Is for Anarchy, V Is for Vendetta: Images of Guy Fawkes and the Creation of Postmodern Anarchism." *Anarchist Studies* 16 (2): 154–72.

camdefendeducation, dir. 2010. *Prof. Richard Drayton on "Economic Lies & Cuts" (1/4).* Economic Lies and Cuts. https://www.youtube.com/watch?v=ymwp5msBPmw.

Candea, Matei. 2019. *Comparison in Anthropology: The Impossible Method.* New Departures in Anthropology. Cambridge University Press.

Captain SKA, dir. 2010. *Liar Liar.* https://www.youtube.com/watch?v=BQFwxw57NBI.

Castells, Manuel. 2012. *Networks of Outrage and Hope: Social Movements in the Internet Age.* Polity.

Chapman, S. D. 1997. "Industry and Trade 1750–1900." In *A Centenary History of Nottingham,* edited by John Beckett, 317–50. Manchester University Press.

Chouliaraki, Lilie. 2006. *The Spectatorship of Suffering.* Sage.

Chua, Liana, and Omri Grinberg. 2021. "Introduction." *Cambridge Journal of Anthropology* 39 (1): 1–17. https://doi.org/10.3167/cja.2021.390102.

Clark, Gregory. 2014. *The Son Also Rises: Surnames and the History of Social Mobility.* Princeton University Press.

Cockerell, Michael, dir. 2016. "Inside the Commons." BBC TWO. https://www.bbc.co.uk/programmes/b05234h3.

Cockroft, Steph. 2014. "Chaos Breaks out in London as Russell Brand Joins Thousands of Masked Guy Fawkes Protesters in Dramatic Bonfire Night Demonstration." *Mail Online.* 5 November 2014. https://www.dailymail.co.uk/news/article-2822591/Chaos-breaks-London-Russell-Brand-joins-thousands-masked-Guy-Fawkes-protesters-dramatic-Bonfire-Night-demonstration.html.

Coleman, E. Gabriella, and Alex Golub. 2008. "Hacker Practice: Moral Genres and the Cultural Articulation of Liberalism." *Anthropological Theory* 8 (3): 255–77. https://doi.org/10.1177/1463499608093814.

Coleman, Gabriella. 2012. 'Our Weirdness Is Free'. *Triple Canopy,* January 2012. https://www.canopycanopycanopy.com/contents/our_weirdness_is_free.

——. 2014. *Hacker, Hoaxer, Whistleblower, Spy: The Many Faces of Anonymous*. Verso Books.

Condorcet, Antoine-Nicolas de. 1955. *Sketch for a Historical Picture of the Progress of the Human Mind*. Translated by June Barraclough. Library of Ideas. Weidenfeld & Nicolson.

Conway, Maura. 2017. "Determining the Role of the Internet in Violent Extremism and Terrorism: Six Suggestions for Progressing Research." *Studies in Conflict & Terrorism* 40 (1): 77–98. https://doi.org/10.1080/1057610X.2016.1157408.

Costa, Elisabetta, Patricia G. Lange, Nell Haynes, and Jolynna Sinanan. 2022. *The Routledge Companion to Media Anthropology*. Routledge Anthropology Handbooks. Routledge.

Cottee, Simon. 2022. *Watching Murder: ISIS, Death Videos and Radicalization*. 1st ed. Routledge.

Crawford, Tom, dir. 2014. *Eviction, The Fraud of the Banks (1 of 3)*. https://www.youtube.com/watch?v=WedSGlpSSO8.

Cressy, David. 1989. *Bonfires and Bells: National Memory and the Protestant Calendar in Elizabethan and Stuart England*. N. Weidenfeld and Nicolson.

——. 1992. "The Fifth of November Remembered." In *Myths of the English*, edited by Roy Porter, 68–90. Polity.

Crewe, Emma. 2015. *The House of Commons: An Anthropology of MPs at Work*. Bloomsbury Academic.

Crumrine, Ross N. 1983. "Masks, Participants and Audience." In *The Power of Symbols: Masks and Masquerade in the Americas*, edited by Ross N. Crumrine and Marjorie M. Halpin, 1–11. University of British Columbia Press.

Dave, Naisargi N. 2014. "Witness: Humans, Animals, and the Politics of Becoming." *Cultural Anthropology* 29 (3): 433–56. https://doi.org/10.14506/ca29.3.01.

Dawson, Lorne L. 1998. *Comprehending Cults: The Sociology of New Religious Movements*. Oxford University Press.

de Heusch, Luc. 1985. *Sacrifice in Africa: A Structuralist Approach*. Translated by Linda O'Brien and Alice Morton. African Systems of Thought. Indiana University Press.

Delfgaauw, Bernard. 1969. *Evolution: The Theory of Teilhard de Chardin*. Translated by Hubert Hoskins. The Fontana Library: Theology and Philosophy. Collins.

Department for Work and Pensions. 2010. "Explanatory Memorandum to the Employment and Support Allowance (Transitional Provisions, Housing Benefit and Council Tax Benefit) (Existing Awards) Regulations 2010." 875. Parliament.

Deseriis, Marco. 2015. *Improper Names: Collective Pseudonyms from the Luddites to Anonymous*. 1st ed. A Quadrant Book Ser. University of Minnesota Press.

Dice, Mark. 2005. *The Resistance Manifesto*. The Resistance.

Diefenbach, Thomas, and John A. A. Sillince. 2011. "Formal and Informal Hierarchy in Different Types of Organization." *Organization Studies* 32 (11): 1515–37. https://doi.org/10.1177/0170840611421254.

Don. 2014. "#OpSafeWinter." In *Know Your Meme*. https://knowyourmeme.com/memes/events/opsafewinter.

Droogers, André. 1980. *The Dangerous Journey: Symbolic Aspects of Boys' Initiation among the Wagenia of Kisangani, Zaire*. Change and Continuity in Africa. Mouton.

Duffy, Simon B. 2013. "A Fair Society? How the Cuts Target Disabled People." Sheffield: Centre for Welfare Reform. https://www.semanticscholar.org/paper/A-fair-society-How-the-cuts-target-disabled-people-Duffy/2f104a9da2c3d7eb9890905fb8e5d5dbdbb6186e.

Dulles, Avery. 1992. *Models of Revelation*. Orbis.

Dumont, Louis. 1986. *Essays on Individualism: Modern Ideology in Anthropological Perspective*. University of Chicago Press.

Eliade, Mircea. 1955. *The Myth of the Eternal Return*. Routledge & Kegan Paul.

——. 1995. *Rites and Symbols of Initiation: The Mysteries of Birth and Rebirth*. Translated by Willard R. Trask. Spring Publications.

Elias, Norbert. 1996. *The Germans: Power Struggles and the Development of Habitus in the Nineteenth and Twentieth Centuries*. Edited by Michael Schröter. Translated by Eric Dunning and Stephen Mennell. Polity Press.

Elkins, Caroline. 2005. *Imperial Reckoning: The Untold Story of Britain's Gulag in Kenya*. Owl Books.

Emigh, John. 1996. *Masked Performance: The Play of Self and Other in Ritual and Theater*. University of Pennsylvania Press.

Errington, Frederick. 1974. *Karavar: Masks and Power in a Melanesian Ritual*. Symbol, Myth, and Ritual Series. Cornell University Press.

Etherington, J. E. 2001. *Lewes Bonfire Night: A Short History of the Guy Fawkes Celebrations*. [Rev. reprint]. Seaford: S. B. Publications.

Fagg, William Buller. 1967. *The Art of Central Africa: Tribal Masks and Sculptures*. Fontana UNESCO Art Books; U23. Collins in association with UNESCO.

Fanon, Frantz. 1970. *Black Skin, White Masks*. Paladin.

Favero, Paolo S. H., and Eva Theunissen. 2018. "With the Smartphone as Field Assistant: Designing, Making, and Testing EthnoAlly, a Multimodal Tool for Conducting Serendipitous Ethnography in a Multisensory World." *American Anthropologist* 120 (1): 163–67. https://doi.org/10.1111/aman.12999.

Ferguson, Marilyn. 1982. *The Aquarian Conspiracy: Personal and Social Transformation in the 1980s*. Granada.

Firer-Blaess, Sylvain. 2016. "The Collective Identity of Anonymous: Web of Meanings in a Digitally Enabled Movement." Uppsala Universitet.

Fisher, Max. 2022. *The Chaos Machine: The Inside Story of How Social Media Rewired Our Minds and Our World*. Little, Brown.

Fletcher, Del Roy, and Sharon Wright. 2018. "A Hand Up or a Slap Down? Criminalising Benefit Claimants in Britain via Strategies of Surveillance, Sanctions and Deterrence." *Critical Social Policy* 38 (2): 323–44. https://doi.org/10.1177/0261018317726622.

Franklin, Sarah. 2019. "Nostalgic Nationalism: How a Discourse of Sacrificial Reproduction Helped Fuel Brexit Britain." *Cultural Anthropology* 34 (1): 41–52. https://doi.org/10.14506/ca34.1.07.

Frazer, James George. 1998. *The Golden Bough: A Study in Magic and Religion*. Oxford World's Classics. Oxford University Press.

Freeman, Jo. 2013. "The Tyranny of Structurelessness." *WSQ: Women's Studies Quarterly* 41 (3): 231–46. https://doi.org/10.1353/wsq.2013.0072.

Freud, Sigmund. 1954. *The Interpretation of Dreams*. Translated by James Strachey. George Allen & Unwin.

Frois, Catarina. 2009. *The Anonymous Society: Identity, Transformation and Anonymity in 12 Step Associations*. Cambridge Scholars.

Fromm, Erich. 1957. *The Forgotten Language: An Introduction to the Understanding of Dreams, Fairy Tales, and Myths*. Evergreen Books, E47. Grove Press.

FTMRecords, dir. 2011. *Commonly Known as Dom—Lawfull Rebellion Workshop @ OccupyLSX*. London. https://www.youtube.com/watch?v=TplPNXTqpg4.

Gehl, Robert. 2009. "YouTube as Archive: Who Will Curate This Digital Wunderkammer?" *International Journal of Cultural Studies* 12 (1): 43–60. https://doi.org/10.1177/1367877908098854.

Geismar, Haidy, and Hannah Knox, eds. 2021. *Digital Anthropology*. 2nd ed. Routledge. https://doi.org/10.4324/9781003087885.

Gell, Alfred. 1975. *Metamorphosis of the Cassowaries: Umeda Society, Language and Ritual*. Monographs on Social Anthropology, no. 51. Athlone Press.

———. 1998. *Art and Agency: An Anthropological Theory*. Clarendon.

Gentleman, Amelia. 2015. "After Hated Atos Quits, Will Maximus Make Work Assessments Less Arduous?" *The Guardian*, 18 January 2015, sec. Society. http://www.theguardian.com/society/2015/jan/18/after-hated-atos-quits-will-maximus-make-work-assessments-less-arduous.

Gesch, Patrick F. 1985. *Initiative and Initiation: A Cargo Cult-Type Movement in the Sepik against Its Background in Traditional Village Religion*. Studia Instituti Anthropos; v. 33. Anthropos-Institut.

Gibson, William. 1995. *Neuromancer*. Voyager.

Gilbert, David, and Felix Driver. 2000. "Capital and Empire: Geographies of Imperial London." *GeoJournal* 51 (1): 23–32. https://doi.org/10.1023/A:1010897127759.

Gluckman, Max. 1962. "Les Rites de Passage." In *Essays on the Ritual of Social Relations*, edited by Daryll Forde, 1–62. Manchester University Press.

Goffman, Erving. 1962. *Asylums: Essays on the Social Situation of Mental Patients and Other Inmates*. Aldine.

Goody, Jack. 2000. "Myth and Masks in West Africa." *Cambridge Anthropology* 22 (2): 60–69.

Graeber, David. 2013. *The Democracy Project: A History, a Crisis, a Movement*. 1st ed. Spiegel & Grau.

Granovetter, Mark S. 1973. "The Strength of Weak Ties." *American Journal of Sociology* 78 (6): 1360–80. https://doi.org/10.1086/225469.

Grant, Mike. 2007. "The Cake Is a Lie." In *Urban Dictionary*. https://www.urbandictionary.com/define.php?term=the%20cake%20is%20a%20lie.

Gray, Jonathan, Liliana Bounegru, and Tommaso Venturini. 2020. "'Fake News' as Infrastructural Uncanny." *New Media & Society* 22 (2): 317–41. https://doi.org/10.1177/1461444819856912.

Grierson, Jamie. 2014. "Huge Police Presence for Anonymous 'Million Mask March.'" *The Mirror*. 5 November 2014. http://www.mirror.co.uk/news/uk-news/anonymous-million-mask-march-huge-4575573.

Handelman, Don. 1990. *Models and Mirrors: Towards an Anthropology of Public Events*. Cambridge University Press.

———. 1993. "Is Victor Turner Receiving His Intellectual Due?" Edited by Bobby C. Alexander and Kathleen M. Ashley. *Journal of Ritual Studies* 7 (2): 117–24.

———. 2021. *Moebius Anthropology: Essays on the Forming of Form*. Edited by Matan Shapiro and Jackie Feldman. Berghahn Books.

Harley, Nicola. 2014. "Masked Protesters Fire Fireworks at Houses of Parliament." *The Telegraph*. 6 November 2014. https://www.telegraph.co.uk/news/uknews/law-and-order/11211981/Masked-protestors-fire-fireworks-at-the-Houses-of-Parliament.html.

Hartwell, Clare. 2002. *Manchester*. Pevsner Buildings of England Series. Yale University Press.

Harvey, David. 2003. *The New Imperialism*. Oxford University Press.

Haugerud, Angelique. 2013. *No Billionaire Left Behind: Satirical Activism in America*. Stanford University Press.

Henderson, Joseph L. 1967. *Thresholds of Initiation*. Middletown, Conn.: Wesleyan University Press.

Herle, Anita, and Jude Philp, eds. 2020. *Recording Kastom: Alfred Haddon's Journals from the Torres Strait and New Quinea, 1888 and 1898*. Indigenous Music of Australia. Sydney University Press.

Hine, Christine. 2015. *Ethnography for the Internet: Embedded, Embodied and Everyday*. Bloomsbury Academic.

Horst, Heather A., and Daniel Miller, eds. 2013. *Digital Anthropology*. Bloomsbury Academic.

——. 2020. *Digital Anthropology*. Routledge, Taylor & Francis Group. https://cam.ldls.org.uk/vdc_100102466447.0x000001.

Howson, Jack. 2014. "Manchester City Council Cuts: Mental Health, Homeless and Children in Care Targeted as Options in Clawback of £59m." *Mancunian Matters*, 12 November 2014. https://www.mancunianmatters.co.uk/news/12112014-manchester-city-council-cuts-mental-health-homeless-and-children-in-care-targeted-as-options-in-clawback-of-59m/.

Hubbard, Barbara Marx. 2015. *Conscious Evolution: Awakening the Power of Our Social Potential*. Revised edition. New World Library.

Hubert, Henri, and Marcel Mauss. 1968. *Sacrifice: Its Nature and Function*. Translated by W. D. Halls. Cohen & West.

Hugh-Jones, Stephen. 1979. *The Palm and the Pleiades: Initiation and Cosmology in Northwest Amazonia*. Cambridge Studies in Social Anthropology 24. Cambridge University Press.

Hutton, Ronald. 1996. *The Stations of the Sun: A History of the Ritual Year in Britain*. Oxford University Press.

Irvine, Richard. 2020. *An Anthropology of Deep Time: Geological Temporality and Social Life*. New Departures in Anthropology. Cambridge University Press.

Jamieson, Mark. 2007. "Masks and Madness. Ritual Expressions of the Transition to Adulthood among Miskitu Adolescents." *Social Anthropology* 9 (3): 257–72. https://doi.org/10.1111/j.1469-8676.2001.tb00152.x.

Jarvie, I. C. 1967. *The Revolution in Anthropology*. Routledge & Kegan Paul.

Jiménez, Alberto Corsín, and Adolfo Estalella. 2023. *Free Culture and the City: Hackers, Commoners, and Neighbors in Madrid, 1997–2017*. Expertise: Cultures and Technologies of Knowledge. Cornell University Press.

Jonaitis, Aldona. 2006. "Sacred Art and Spiritual Power: An Analysis of Tlingit Shamans' Masks." In *The Anthropology of Art: A Reader*, edited by Morgan Perkins and Howard Morphy, 358–73. Blackwell Anthologies in Social and Cultural Anthropology. Blackwell.

Jones, Alex, dir. 2007. *Endgame: Blueprint for Global Enslavement*. Documentary. Disinformation Company.

Joseph, Francis, and Mary Bonongwe. 2005. "Initiation Rites for Boys in Lomwe Society: A Dying Cultural Practice." In *Initiation Rites for Boys in Lomwe Society in Malawi and Other Essays*, edited by J. C. Chakanza, 9–38. Sources for the Study of Religion in Malawi, no. 23. Zomba: Kachere Series.

Jung, C. G. 1953. *Two Essays on Analytical Psychology*. Translated by R. F. C. Hull. The Collected Works of C. G. Jung, v. 7. Routledge & Kegan Paul.

Juris, Jeffrey S. 2008. *Networking Futures: The Movements against Corporate Globalization*. Duke University Press.

Kalb, Don. 2018. "Austerity and 'the Discipline of Historical Context'." In *The Global Life of Austerity: Comparing Beyond Europe*, edited by Theodoros Rakopoulos, 130–50. Critical Interventions: A Forum for Social Analysis 17. New York: Berghahn Books.

Kavoori, Anandam P. 2011. *Reading YouTube: The Critical Viewers Guide*. Digital Formations, v. 64. Peter Lang.

Kent, Stephen A. 2015. "Freemen, Sovereign Citizens, and the Challenge to Public Order in British Heritage Countries." *International Journal of Cultic Studies* 6:1–16.

Kersey, John. 2010. "The Freemen on the Land Movement: Grass Roots Libertarianism in Action." Legal Notes 50. The Libertarian Alliance. http://libertarian. co.uk/2021/08/17/legal-notes-50-the-freemen-of-the-land-movement-grass-roots-libertarianism-in-action-2010-by-john-kersey/.

Khomami, Nadia. 2017. "Cambridge Student Filmed Burning Cash in Front of Homeless Man." *The Guardian*, 10 February 2017, sec. UK News. https://www. theguardian.com/uk-news/2017/feb/10/cambridge-student-filmed-burning-money-in-front-of-homeless-man.

Kivanç, Jake. 2016. "We Spoke to a Leader of Canada's Freemen on the Land Movement about the Oregon Standoff." *VICE* (blog). 8 January 2016. https:// www.vice.com/en/article/we-spoke-to-a-leader-in-the-freemen-on-the-land-movement-about-the-oregon-standoff/.

Knappenberger, Brian, dir. 2012. *We Are Legion: The Story of the Hacktivists*. FilmBuff. https://en.wikipedia.org/w/index.php?title=We_Are_Legion&oldid=666443478.

Knight, Stephen. 2003. *Robin Hood: A Mythic Biography*. Cornell University Press.

Koloss, Hans-Joachim. 1988. "Kings, Masks, and Festivals in the Grasslands of Cameroon." *Visual Anthropology* 1 (3): 287–91. https://doi.org/10.1080/0894946 8.1988.9966483.

Konzelmann, Suzanne J. 2019. *Austerity*. What Is Political Economy? Polity.

Kruglova, Anna. 2022. *Terrorist Recruitment, Propaganda and Branding: Selling Terror Online*. 1st ed. Political Violence. Routledge. https://cam.ldls.org.uk/ vdc_100156088401.0x000001.

La Fontaine, J. S. 1985. *Initiation*. Themes in Social Anthropology. Penguin Books.

Labrum, E. A. 1994. *Civil Engineering Heritage. Eastern and Central England*. London: Published for the Institution of Civil Engineers by Thomas Telford.

Laidlaw, James. 2000. "A Free Gift Makes No Friends." *Journal of the Royal Anthropological Institute* 6 (4): 617–34. https://doi.org/10.1111/1467-9655.00036.

Lane, David H. 1996. *The Phenomenon of Teilhard: Prophet for a New Age*. Mercer University Press.

Lange, Patricia G. 2016. *Kids on YouTube: Technical Identities and Digital Literacies*. 1st ed. Routledge.

——. 2019. *Thanks for Watching: An Anthropological Study of Video Sharing on YouTube*. University Press of Colorado.

Lawrence, Peter. 1964. *Road Belong Cargo: A Study of the Cargo Movement in the Southern Madang District, New Guinea*. Manchester University Press.

Lazar, Sian. 2018. "A 'Kinship Anthropology of Polities'? Interest, the Collective Self, and Kinship in Argentine Unions." *The Journal of the Royal Anthropological Institute* 24 (2): 256–74.

Lévi-Strauss, Claude. 1983. *The Way of the Masks*. Cape.

Library of Congress. n.d. "Freeman-on-the-Land Movement." LC Linked Data Service: Authorities and Vocabularies. Webpage. Accessed 5 November 2024. https:// id.loc.gov/authorities/subjects/sh2013000336.html.

Lincoln, Bruce. 1981. *Emerging from the Chrysalis: Studies in Rituals of Women's Initiation*. Harvard University Press.

Lindstrom, Lamont. 1993. *Cargo Cult: Strange Stories of Desire from Melanesia and Beyond*. South Sea Books. University of Hawaii Press.

Littler, Mark, and Benjamin Lee, eds. 2020. *Digital Extremisms: Readings in Violence, Radicalisation and Extremism in the Online Space*. Palgrave Studies in Cybercrime and Cybersecurity. Palgrave Macmillan. https://cam.ldls.org.uk/vdc_100096692685.0x000001.

Lowe, David, and Jack Richards. 1982. *The City of Lace*. Nottingham Lace Centre.

Margetts, Helen, Peter John, Scott A. Hale, and Taha Yasseri. 2016. *Political Turbulence: How Social Media Shape Collective Action*. Princeton University Press.

Mark, Peter. 1992. *The Wild Bull and the Sacred Forest: Form, Meaning, and Change in Senegambian Initiation Masks*. Cambridge University Press.

Markman, Peter T., and Roberta H. Markman. 1989. *Masks of the Spirit: Image and Metaphor in Mesoamerica*. University of California Press.

Martino, John. 2018. *Drumbeat: New Media and the Radicalization and Militarization of Young People* 1st ed. London: Routledge.

Marwick, Alice E., and Becca Lewis. 2017. "Media Manipulation and Disinformation Online." Data & Society Research Institute. https://datasociety.net/library/media-manipulation-and-disinfo-online/.

Marx, Karl. 1995. *Capital: An Abridged Edition*. Oxford World's Classics. Oxford University Press.

Mauss, Marcel. 1985. "A Category of the Human Mind: The Notion of Person; the Notion of Self." In *The Category of the Person: Anthropology, Philosophy, History*, edited by Michael Carrithers, Steven Collins, and Steven Lukes, 1–25. Cambridge University Press.

——. 2016. *The Gift*. Translated by Jane I Guyer. Expanded edition. HAU Books.

Mayblin, Maya, and Magnus Course. 2014. "The Other Side of Sacrifice: Introduction." *Ethnos* 79 (3): 307–19. https://doi.org/10.1080/00141844.2013.841720.

McGovern, Mike. 2013. *Unmasking the State: Making Guinea Modern*. University of Chicago Press.

McGrain, Pete, dir. 2011. *Ethos*. Documentary. Media for Action.

McKernan, Luke. 2016. "The Disappearing Archive." *Luke McKernan* (blog). 3 January 2016. https://lukemckernan.com/2016/01/03/the-disappearing-archive/.

McTeigue, James, dir. 2005. *V for Vendetta*. Warner Bros. Pictures.

Mead, Margaret. 1956. *New Lives for Old: Cultural Transformation—Manus, 1928–1953*. Gollancz.

Mendax, Neo, dir. 2014. *STOP the Eviction*. https://www.youtube.com/watch?v=EvVNANqQY6U.

Miller, Daniel, and Heather A. Horst. 2012. "The Digital and the Human: A Prospectus for Digital Anthropology." In *Digital Anthropology*, edited by Heather A. Horst and Daniel Miller. Routledge.

Miller, Daniel, and Don Slater. 2003. *The Internet: An Ethnographic Approach*. Berg.

mind open, dir. 2014. *#OpSafeWinter—Nottingham–13th December 2014*. https://www.youtube.com/watch?v=3gdqFuLNCLs.

Moore, Alan. 2008. *V for Vendetta*. New edition. Vertigo.

Müller, Jan-Werner. 2020. "Why Do Rightwing Populist Leaders Oppose Experts?" *The Guardian*, 26 March 2020, sec. Opinion. https://www.theguardian.com/commentisfree/2020/mar/26/rightwing-populist-leaders-oppose-experts-not-elites.

Munn, Luke. 2019. "Alt-Right Pipeline: Individual Journeys to Extremism Online." *First Monday* 24 (June). https://doi.org/10.5210/fm.v24i6.10108.

——. 2023. *Red Pilled: The Allure of Digital Hate*. BiUP General. Bielefeld University Press, 2023.

Myers, Fred. 2017. "Rant or Reason: Old Wine and New Bottles in Anthropology." *HAU: Journal of Ethnographic Theory* 7 (January): 8–12.

Nagel, Thomas. 1986. *The View from Nowhere*. Oxford University Press.

Nagle, Angela. 2017. *Kill All Normies: The Online Culture Wars from Tumblr and 4chan to the Alt-Right and Trump*. Zero Books.

Napier, A. David. 1986. *Masks, Transformation, and Paradox*. University of California Press.

Neal, Andrew, Sven Opitz, and Chris Zebrowski. 2019. "Capturing Protest in Urban Environments: The 'Police Kettle' as a Territorial Strategy." *Environment and Planning D: Society and Space* 37 (6): 1045–63. https://doi.org/10.1177/0263775819841912.

Nippert-Eng, Christena E. 1996. *Home and Work: Negotiating Boundaries through Everyday Life*. University of Chicago Press.

——. 2005. "Boundary Play." *Space and Culture* 8 (3): 302–24. https://doi.org/10.1177/1206331205277351.

Odell, Jenny. 2020. *How to Do Nothing: Resisting the Attention Economy*. Melville House.

Ogle, Vanessa. 2015. *The Global Transformation of Time: 1870–1950*. Harvard University Press.

O'Hara, Mary. 2015. *Austerity Bites: A Journey to the Sharp End of Cuts in the UK*. Bristol, UK: Policy Press.

Oliver, Kelly. 2001. *Witnessing: Beyond Recognition*. University of Minnesota Press.

Olson, Parmy. 2013. *We Are Anonymous: Inside the Hacker World of LulzSec, Anonymous, and the Global Cyber Insurgency*. Back Bay Books.

Oosten, Jarich. 1992. "Representing the Spirits: The Masks of the Alaskan Inuit." In *Anthropology, Art, and Aesthetics*, edited by Jeremy Coote and Anthony Shelton, 113–34. Oxford Studies in the Anthropology of Cultural Forms. Clarendon.

Osborne, John W. 1966. *William Cobbett: His Thought and His Times*. Rutgers University Press.

Pariser, Eli. 2012. *The Filter Bubble: What the Internet Is Hiding from You*. Penguin Books.

Parveen, Nazia. 2015. "Bailiffs Trying to Evict Cancer-Stricken Man Retreat." *Mail Online*, 23 January 2015, sec. News. https://www.dailymail.co.uk/news/article-2923567/Bailiffs-trying-evict-cancer-stricken-father-mortgage-blunder-forced-retreat-500-strangers-form-human-blockade-bungalow.html.

Patterson, Orlando. 1982. *Slavery and Social Death: A Comparative Study*. Harvard University Press.

Peacock, Vita. 2007. "The Politics and Transculturation of British Popular Music, circa 1976–1981." BA Thesis: University of Cambridge.

——. 2025. 'Scatological Humour and the Absent Anthropology of Privacy'. *Surveillance & Society* 23 (1): 101–11. https://doi.org/10.24908/ss.v23i1.16827.

Peters, John Durham. 2001. "Witnessing." *Media, Culture & Society* 23 (6): 707–23. https://doi.org/10.1177/016344301023006002.

Pink, Sarah, Heather A. Horst, John Postill, Larissa Hjorth, Tania Lewis, and Jo Tacchi. 2016. *Digital Ethnography: Principles and Practice*. SAGE.

Piven, Frances Fox, and Richard A. Cloward. 1977. *Poor People's Movements: Why They Succeed, How They Fail*. Pantheon Books.

Playford, Alison, Neil Howard, and commonly known as Dom. 2011. "We Are the Change: Welfare, Education and Law at the Occupy Camp." *The Guardian*, 15 November 2011, sec. Opinion. https://www.theguardian.com/commentisfree/2011/nov/15/welfare-education-law-occupy-london.

Poole, F. J. P. 1982. "The Ritual Forging of Identity: Aspects of Person and Self in Bimin-Kuskusmin Male Initiation." In *Rituals of Manhood: Male Initiation in Papua New Guinea*, edited by Gilbert H. Herdt. University of California Press.

Popper, Karl R. 1959. *The Logic of Scientific Discovery*. Hutchinson, Basic Books.

Postill, John. 2012. "Facebook for Ethnographers: Data-rich, Distracting, and Awkward." *Media/Anthropology* (blog). 17 September 2012. https://johnpostill.wordpress.com/2012/09/17/facebook-for-ethnographers-data-%c2%adrich-distracting-and-awkward/.

——. 2014. "Democracy in an Age of Viral Reality: A Media Epidemiography of Spain's Indignados Movement." *Ethnography* 15 (1): 51–69. https://doi.org/10.1177/1466138113502513.

Procházka, Ondřej, and Jan Blommaert. 2021. "Ergoic Framing in New Right Online Groups: Q, the MAGA Kid, and the Deep State Theory." *Australian Review of Applied Linguistics* 44 (1): 4–36. https://doi.org/10.1075/aral.19033.pro.

Quays News. 2015. "The Homeless in Manchester—Protestors or Communities?" *Quays News* (blog). 19 October 2015. http://quaysnews.net/index.php/2015/10/19/the-homeless-in-manchester-protestors-or-communities/.

Quinn, Ben. 2014. "Million Mask March Draws Thousands in London on Global Day of Protest." *The Guardian*, 5 November 2014, sec. UK news. https://www.theguardian.com/uk-news/2014/nov/05/million-mask-march-london-russell-brand-anonymous.

Rakopoulos, Theodoros, ed. 2018. *The Global Life of Austerity: Comparing Beyond Europe*. Critical Interventions: A Forum for Social Analysis 17. Berghahn Books.

Raposo, Paulo. 2005. "Masks, Performance and Tradition: Local Identities and Global Contexts." *Etnográfica: Revista Do Centro Em Rede de Investigação Em Antropologia* 9 (1): 49–64.

Richards, Audrey I. 1988. *Chisungu: A Girls' Initiation Ceremony among the Bemba of Zambia*. Routledge.

Richardson, Allissa V. 2020. *Bearing Witness While Black: African Americans, Smartphones, and the New Protest #Journalism*. Oxford University Press, 2020.

Robbins, Joel. 2004. *Becoming Sinners: Christianity and Moral Torment in a Papua New Guinea Society*. Ethnographic Studies in Subjectivity 4. University of California Press.

Robbins, Thomas. 1988. *Cults, Converts, and Charisma: The Sociology of New Religious Movements*. Sage.

Roberts, Noel. 2000. *From Piltdown Man to Point Omega: The Evolutionary Theory of Teilhard de Chardin*. Studies in European Thought, v. 18. P. Lang.

Robertson, David G. 2016. *UFOs, Conspiracy Theories and the New Age: Millennial Conspiracism*. Bloomsbury Advances in Religious Studies. Bloomsbury Academic.

Robertson, Pat. 1991. *The New World Order*. Word.

Rucki, Alexandra, and Ben Morgan. 2014. "Million Mask March: Protesters Arrested as Violence Breaks out in Central London." *The Standard*. 10 November 2014. https://www.standard.co.uk/news/london/fireworks-and-flares-let-off-as-million-mask-march-descends-on-central-london-9842218.html.

Runton, P. T. 1942. *The Key of Masonic Initiation*. J. M. Watkins.

Sanchez, Andrew. 2023. "Kill Your Ancestors." *American Ethnologist* 50 (3): 439–45. https://doi.org/10.1111/amet.13190.

Saul, Heather. 2014. "Million Mask March Protesters Attack Woman 'for Having the Wrong Accent'." *The Independent*. 7 November 2014. https://www.independent.co.uk/news/uk/home-news/

million-mask-march-protesters-attack-woman-for-having-the-wrong-accent-9844085.html.

Schulz, Kathryn. 2015. "The Rabbit-Hole Rabbit Hole." *The New Yorker*, 4 June 2015. https://www.newyorker.com/culture/cultural-comment/the-rabbit-hole-rabbit-hole.

Scobie, Jon, dir. 2011. *John Harris—It's an Illusion*. https://www.youtube.com/watch?v=NKvacmyLG2A.

Shapin, Steven, and Simon Schaffer. 1985. *Leviathan and the Air-Pump: Hobbes, Boyle, and the Experimental Life*. Princeton University Press.

Shea, Matt, dir. 2023. *Andrew Tate: The Man Who Groomed the World?* Documentary. https://www.bbc.co.uk/programmes/m001q1n6.

Simmel, Georg. 1906. "The Sociology of Secrecy and of Secret Societies." *American Journal of Sociology*, January, 441–98.

Smith, Pierre. 1982. "Aspects of the Organization of Rites." In *Between Belief and Transgression: Structuralist Essays in Religion, History, and Myth*, edited by Michel Izard and Pierre Smith, translated by John Leavitt, 103–28. Chicago Originals. University of Chicago Press.

Snickars, Pelle, and Patrick Vonderau. 2009. *The YouTube Reader*. Mediehistoriskt Arkiv 12. National Library of Sweden.

Sparrow, Andrew. 2013. "Margaret Thatcher's Funeral Cost Taxpayers More than £3m." *The Guardian*, 29 July 2013, sec. Politics. https://www.theguardian.com/politics/2013/jul/29/margaret-thatcher-funeral-cost-taxpayers.

Srnicek, Nick. 2016. "Platform Capitalism." In *Platform Capitalism*, 36–92. Polity Press.

Stack, Liam, Nick Cumming-Bruce, and Madeleine Kruhly. 2019. "How Julian Assange and WikiLeaks Became Targets of the U.S. Government." *The New York Times*, 11 April 2019, sec. World. https://www.nytimes.com/interactive/2019/world/julian-assange-wikileaks.html, https://www.nytimes.com/interactive/2019/world/julian-assange-wikileaks.html.

Stark, Rodney, and William Sims Bainbridge. 1985. *The Future of Religion: Secularization, Revival, and Cult Formation*. University of California Press.

Steinbauer, Friedrich. 1979. *Melanesian Cargo Cults: New Salvation Movements in the South Pacific*. Translated by Max Wohlwill. G. Prior.

Steiner, Christopher B. 1992. "The Invisible Face: Masks, Ethnicity, and the State in Côte d'Ivoire, West Africa." *Museum Anthropology* 16 (3): 53–57. https://doi.org/10.1525/mua.1992.16.3.53.

Steiner, Rudolf. 1991. *Old and New Methods of Initiation: Fourteen Lectures given in Dornach, Mannheim and Breslau from 1 January to 19 March 1922*. Translated by Johanna Collis. Rudolf Steiner Press.

Stewart, Kathleen, and Susan Harding. 1999. "Bad Endings: American Apocalypsis." *Annual Review of Anthropology* 28: 285–310. https://doi.org/10.1146/annurev.anthro.28.1.285.

Stone, Oliver, dir. 1987. *Wall Street*. 20th Century Fox.

Strathern, Andrew. 1975. *The Rope of Moka: Big-Men and Ceremonial Exchange in Mount Hagen New Guinea*. Cambridge Studies in Social and Cultural Anthropology No. 4. Cambridge University Press.

Sullivan, Randall. 2009. "American Stonehenge: Monumental Instructions for the Post-Apocalypse." *WIRED*, 20 April 2009. https://www.wired.com/2009/04/ff-guidestones/.

Swartz, Aaron. 2016. *The Boy Who Could Change the World: The Writings of Aaron Swartz*. Verso.

Taussig, Michael T. 1999. *Defacement: Public Secrecy and the Labor of the Negative*. Stanford University Press.

——. 2012. "The Stories Things Tell And Why They Tell Them." *E-Flux Journal*, no. #36. Accessed 19 November 2024. https://www.e-flux.com/journal/36/61256/the-stories-things-tell-and-why-they-tell-them/.

Tebbs, H. F. 1979. *Peterborough: A History*. Cambridge Town, Gown and County 24. Oleander Press.

The Newsroom. 2017. "Amazon Creates 2,500 Seasonal Jobs in Peterborough." *Peterborough Telegraph*, 3 October 2017. https://www.peterboroughtoday.co.uk/business/amazon-creates-2500-seasonal-jobs-in-peterborough-1076624.

The Prince of Diamonds. 2012. "Twerking." In *Urban Dictionary*. https://www.urbandictionary.com/define.php?term=Twerking&page=5.

The Telegraph. 2015. "Hundreds of Strangers Form Blockade to Help Man Claim Eviction Victory over Bailiffs." *The Telegraph*, 23 January 2015. https://www.telegraph.co.uk/news/uknews/law-and-order/11365475/Mans-victory-over-bailiffs-thanks-to-500-strangers-who-stop-eviction.html.

Thompson, E. P. 1967. "Time, Work-Discipline, and Industrial Capitalism." *Past & Present*, no. 38, 56–97.

Titmuss, Richard Morris. 2002. *The Gift Relationship: From Human Blood to Social Policy*. Titmuss, Richard Morris, 1907–1973. Writings on Social Policy and Welfare. Palgrave Macmillan Archive Ed. 2002 7. Palgrave Macmillan.

Tolle, Eckhart. 1999. *The Power of Now: A Guide to Spiritual Enlightenment*. Hodder and Stoughton.

Tonkin, Elizabeth. 1979. "Masks and Powers." *Man* 14 (2): 237–48. https://doi.org/10.2307/2801565.

Tooze, Adam. 2019. *Crashed: How a Decade of Financial Crises Changed the World*. Penguin Books.

Tourish, Dennis, and Tim Wohlforth. 2000. *On the Edge: Political Cults Right and Left*. M. E. Sharpe.

Troeltsch, Ernst. 1981. *The Social Teaching of the Christian Churches*. University of Chicago Press.

Turner, Bryan S. 1986. "Personhood and Citizenship." *Theory, Culture & Society* 3 (1): 1–16. https://doi.org/10.1177/0263276486003001002.

Turner, Victor W. 1967. *The Forest of Symbols; Aspects of Ndembu Ritual*. Cornell University Press.

——. 1969. *The Ritual Process: Structure and Anti-Structure*. The Lewis Henry Morgan Lectures 1966. Aldine.

——. 1974a. *Dramas, Fields, and Metaphors; Symbolic Action in Human Society*. Symbol, Myth, and Ritual. Cornell University Press.

——. 1974b. "Liminal to Liminoid, in Play, Flow, and Ritual: An Essay in Comparative Symbology." *Rice Institute Pamphlet—Rice University Studies* 60 (3). https://scholarship.rice.edu/handle/1911/63159.

——. 1978. "Comments and Conclusions." In *The Reversible World: Symbolic Inversion in Art and Society*, edited by Barbara A. Babcock, 276–96. Symbol, Myth, and Ritual Series. Cornell University Press.

Tuzin. 1982. "The Ritual Forging of Identity: Aspects of Person and Self in Bimin-Kuskusmin Male Initiation." In *Rituals of Manhood: Male Initiation in Papua New Guinea*, edited by Gilbert H. Herdt. University of California Press.

Udupa, Sahana. 2019. "Nationalism in the Digital Age: Fun as a Metapractice of Extreme Speech." *International Journal of Communication*, 3143–63. https://doi.org/10.5282/ubm/epub.69633.

Uscinski, Joseph E., and Joseph Parent. 2014. *American Conspiracy Theories*. Oxford University Press.

Vacca, John R., ed. 2019. *Online Terrorist Propaganda, Recruitment, and Radicalization*. CRC Press. https://doi.org/10.1201/9781315170251.

Valeri, Valerio. 1985. *Kingship and Sacrifice: Ritual and Society in Ancient Hawaii*. University of Chicago Press.

Van Badham, Vanessa. 2021. *QAnon and On: A Short and Shocking History of Internet Conspiracy Cults*. Hardie Grant Books.

Van der Veer, Peter. 2016. *The Value of Comparison*. Lewis Henry Morgan Lectures. Duke University Press.

Van Gennep, Arnold. 1977. *The Rites of Passage*. Routledge & Kegan Paul.

Wallace, Anthony F. C. 1956. "Revitalization Movements." *American Anthropologist* 58 (2): 264–81. https://doi.org/10.2307/665488.

Wallis, William, Anna Gross, and Lucy Fisher. 2024. "Far-Right Groups Plan to Target Homes in Attack on Lawyers, Police Told." *Financial Times*, 6 August 2024, sec. UK Riots. https://www.ft.com/content/4daf5d4e-cdf9-4175-b101-925573b5f021.

Watkins, Johnathan, Wahyu Wulaningsih, Charlie Zhou, Dominic Marshall, Guia Sylianteng, Gia Dela Rosa, Viveka Miguel, Rosalind Raine, Lawrence King, and Mahiben Maruthappu. 2017. "Effects of Health and Social Care Spending Constraints on Mortality in England: A Time Trend Analysis." *BMJ Open* 7 (November): e017722. https://doi.org/10.1136/bmjopen-2017-017722.

Webb, Sam. 2018. "What Is the Bullingdon Club at Oxford University, Why Do They Burn £50 Notes during Initiation Tests and Who Has Been a Member?" *The Sun*, 15 February 2018. https://www.thesun.co.uk/news/2833129/bullingdon-club-oxford-why-burn-50-notes/.

Weber, Max. 1966. *The Sociology of Religion*. Translated by Ephraim Fischoff. Social Science Paperbacks in association with Methuen, Methuen, Methuen & Co.

Webster, Hutton. 1932. *Primitive Secret Societies: A Study in Early Politics and Religion*. 2d ed., rev. Macmillan.

Wesch, Michael, dir. 2008. *An Anthropological Introduction to YouTube*. https://www.youtube.com/watch?v=TPAO-lZ4_hU.

West, Harry G., and Todd Sanders, eds. 2003. *Transparency and Conspiracy: Ethnographies of Suspicion in the New World Order*. Duke University Press.

Williams, James. 2018. *Stand out of Our Light: Freedom and Resistance in the Attention Economy*. Cambridge University Press.

Williams, Jennifer. 2015. "Manchester Town Hall in Lockdown as Anti-Austerity Demonstrators Attempt to Storm Building." *Manchester Evening News*, 15 April 2015, sec. Greater Manchester News. http://www.manchestereveningnews.co.uk/news/greater-manchester-news/manchester-town-hall-lockdown-anti-austerity-9052240.

Wilmshurst, W L. 1924. *The Masonic Initiation*. William Rider & Son, Ltd. and Percy Lund Humphries & Co., Ltd.

Winter, J. C. 2000. *Tobacco Use by Native North Americans: Sacred Smoke and Silent Killer*. University of Oklahoma Press.

Worsley, Peter. 1957. *The Trumpet Shall Sound: A Study of "Cargo" Cults in Melanesia*. MacGibbon & Kee.

Zapatistas, The. 1998. *Zapatista Encuentro: Documents from the First Intercontinental Encounter for Humanity and Against Neoliberalism*. 1st ed. New York: Seven Stories Press.

Zuboff, Shoshana. 2019. *The Age of Surveillance Capitalism: The Fight for a Human Future at the New Frontier of Power*, 1st ed. PublicAffairs.

Index

Page numbers in *italics* refer to figures and tables.

Benjamin, Walter, 51, 117, 125, 184n25
Betty (AnonUK member), 131, 143–46,
    167–69, 192n24
Beyer, Jessica, 7, 179nn49–50
Bez (AnonUK member), 120, 122–23
biblical inerrancy, 97, 110–11, 193n47
Big Ben, 112, 115–17
Bilderbergers, 61, 186n17
birth certificates, 65–69, 85, 90, 187n39,
    191n48
Black Lives Matter movement, 55
Blair, Tony, 107–9
blindfolds, 29, 86–87, 190n38, 195n18
blood sacrifice, 156–57, 197n35
bodily experiences, 24; prophetic visions,
    33–34, 56, 166, 182n17; separation and
    absence of care, 29–41
Bonfire Boys, 5, 177n10
Bonfire Night. See November 5th
boundary play/work, 74–75
Braveheart (film), *47*, 73–74
British government: imperial power of, 2, 116;
    institutions as corporations, 67; national
    debt, 66, 157; as They, 61. *See also* austerity
    politics; Conservative Party; Labour Party;
    welfare policy
British holidays and festivals, 125–29. *See also*
    November 5th; *specific holidays*
Brown, Gordon, 19
Buckingham Palace, 2, 6, 61, 163
Bush, George H.W., 69

calendrical time, 125–29
Call, Lewis, 7, 77
Cameron, David, 20, 61, 157, 198n40
capitalism, 50–52, 98, 129; clock-time and,
    115–17, 123
care practices: failure of (*see* structural
    carelessness); nontransactional, 153; in
    traditional initiation, 39–40
cargo cults of Melanesia and Polynesia, 22, 33
Catholicism, 4, 17, 31, 128, 134, 145, 181n3,
    187n42
charity, 154, 171
Chartist movement, 149
children, 18, 29, 71, 90, 102, 131–32, 139, 142,
    192n19. *See also* puberty rites
Christianity: bearing witness, 44; divine
    revelation and biblical inerrancy, 97,
    110–11, 193n47; in early British history,
    118–19; Indigenous rituals, 113; mysteries
    in, 192n32; Paul's conversion to, 33; post-
    Lutheran, 86; postmillennialism, 131–32,

187n38; premillennialism, 70, 131, 187n38;
    sacrifice and, 157; theology, 72, 97;
    traditional initiation masks and, 90. *See also*
    Catholicism; Salvation Army
Christmas, 125, *127*, 150
citation practices, 53
citizenship, 90–91, 187n29
civil disobedience, 32
civil or statute law, 67
clubs, 11, 87
cognitive experience, 30–31, 41
Coleman, Gabriella, 7, 14, 59, 73, 77, 153,
    179n49, 179n50
commensality, 8, 137, 161
commodity fetish, 120, 129
common law jurisdiction, 67, 69
communitas, 13, 23, 79–80, 164, 167, 178n34,
    180n80
consciousness: altered states of, 41; arrival
    into, 133; knowledge and, 21, 62; New Age
    theories of, 116; planetary, 135. *See also*
    waking up
Conservative Party, 19–20, 109, 138, 150, 158
Conservative-Liberal Democrat coalition, 19,
    182n22
conspiracy theory, 23, 169–70, 180nn82–83.
    See also vernacular theories
corruption, 61, 99–100, 138, 186n22
cosmogony, 43, 52, 106
cosmology, 8, 43, 52, 56, 58, 112, 165; occult,
    58, 72
Course, Magnus, 155, 159–60, 197n27
coverings, 30, 86–87, 181n3, 190n45. *See also*
    blindfolds; masks
Covid-19 (coronavirus) pandemic, 72, 167–68,
    187n41, 187n48
Crawford, Tom, 158–61
cults: defined, 3, 22, 177n7; negative views
    of, 23, 131, 180n81; political, 106; social
    conflict and, 13; world-building and, 112.
    See also online cults
cyberspace, 8, 43, 183n2
cyberutopianism, 14–15, 19, 70, 135, 153, 169

De Chardin, Teilhard, 134–35, 194n9
death, 5, 32–33, 69, 71–72, 78, 121. *See also*
    social death; symbolic death
death masks, 87, 190n40
defacement, 86–87, 190n37, 190n39; of
    symbols, 114–17, 119; virtual, 117
dehumanization, 62, 66
Depopulation Agenda, 21, 46, 58, 66, 71–72,
    103, 166, 169

www.ingramcontent.com/pod-product-compliance
Lightning Source LLC
Chambersburg PA
CBHW030317270326
41926CB00010B/1406